Islands and the Modernists

I0792760

Islands and the Modernists

The Allure of Isolation in Art, Literature and Science

JILL FRANKS

McFarland & Company, Inc., Publishers
Jefferson, North Carolina, and London

Frontispiece: **John Millington Synge in July 1906.**

LIBRARY OF CONGRESS CATALOGUING-IN-PUBLICATION DATA

Franks, Jill, 1957–
Islands and the modernists : the allure of isolation in art, literature and science / Jill Franks.
p. cm.
Includes bibliographical references and index.

ISBN-13: 978-0-7864-2457-3
ISBN-10: 0-7864-2457-5 (softcover : 50# alkaline paper) ♾

1. Modernism (Aesthetics). 2. Islands—Miscellanea.
I. Title.
BH301.M54F73 2006
001.09'034—dc22 2006013453

British Library cataloguing data are available

©2006 Jill Franks. All rights reserved

No part of this book may be reproduced or transmitted in any form or by any means, electronic or mechanical, including photocopying or recording, or by any information storage and retrieval system, without permission in writing from the publisher.

Cover photograph ©2005 Brand X Pictures

Manufactured in the United States of America

McFarland & Company, Inc., Publishers
Box 611, Jefferson, North Carolina 28640
www.mcfarlandpub.com

Acknowledgments

In the course of writing a book on islands, dreams, and isolation, I struggled to avoid my own mental isolation. Interdisciplinary work affords opportunities to meet scholars in different fields. Several people aided me in the endeavor, by reading, discussing, or providing materials. I wrote most of this book away from home, in a beautiful, secluded "island" of a mountain retreat. For this, I would like to thank Charlie Yowell, who provided a cozy study and companionship.

I thank the University of Tennessee at Chattanooga librarian Andrea Anderson for extending borrowing privileges, and Professor Thomas Ware for establishing the scholarly affiliation that made these privileges possible (and for remembering what Melville said about the ocean and horror). I would also like to thank the staff and faculty at Austin Peay State University who helped with the project.

Carol Kimmel has been generous with library privileges, and colleagues David Till, Steve Ryan, and Jim Thompson have read chapters, commenting insightfully on style. Jim also provided guidance in molecular genetics and brilliant pictures of the Galapagos.

My mother, Mary Franks, is an excellent reader who assured me that her own field of art history is as political as my field of literature. I feel indebted to her powers of imagination and witty conversation as inspiration to think about and believe in the project.

I have not lived on a small island in a very long time, but I would like to acknowledge Martha's Vineyard as a source of inspiration for me. Ever since I lived there in the summer of 1977, it has held for me what Dea Birkett calls the perfect place in the heart, which is in the shape of an island.

Table of Contents

Preface

This book is about the centrality of islands to human dreams. Since my academic training is in the field of modern British literature, and because modernism is characterized by dreaming, the desire for escape, and projects for social modification, I readily noticed that islands fulfill a specific need in the human psyche, which I refer to as "imaginable space." Given their bounded-ness and their distance from a mainland, islands suggest themselves as both controllable and paradisiacal. For Charles Darwin and Margaret Mead, the sense of control refers to the careful limiting of their scientific or social laboratory so that hypotheses may be specific enough to be tested thoroughly. For Paul Gauguin and D. H. Lawrence, the fanciful idea of paradise compels, no matter how sophisticated their artists' minds. For John Millington Synge, islands provide a hybrid space between art and science: Aran gives him a language lab where Gaelic is spoken and a rich cultural fabric waiting to be woven into drama for the rest of Ireland to witness. For all five of them, their islands of choice are both literal and imaginative space that function as crucibles for their creativity.

The major modern figures I chose are known for their individualist, and sometimes radical, qualities. Each one pioneered in his or her field. Each of them uses islands differently—from Lawrence, who discovers he does not like physical islands, but sustains himself through a lifetime by what he calls his island scheme, to Mead, who also dislikes isolation, but by her work on islands makes so many contacts that she acquires the sobriquet "Mother of the World." Each of the five also uses islands in similar ways to the others. One of the more exciting aspects of my research has been making interesting connections across the fields and among the five historical figures involved. For each of these five innovators, the balance between isolation and togetherness, between tradition and innovation, between reality and the imagination was difficult to achieve. Interdisciplinary work shares some of the same pitfalls and tensions. In my appeal to an interdisciplinary academic audience, I have had to shift tone, from the theory-bound, jargon-laden one common to English literary studies, to a broader, more informative one. In contrast to literary criticism, I here provide more factual and biographical data, and I do not hesitate to pose psychological suggestions arising from the evidence.

Most single-author books about islands that I have encountered during the five years of my research have been more specific in focus than mine. They have not been interdisciplinary, but written by geographers, biologists, historians, political scientists, and economists. *Islands and the Modernists* blends the four disciplines of evolutionary biology, social anthropology, painting, and literature. The range of disciplines that have begun to take an interest in island studies is amply demonstrated by a glance at some of the paper titles on the program for the "Islands of the World" conference held at the Institute of Island Studies at the University of Prince Edward Island in 2002: "Direct Democracy in Microstates and Small Island States," "Promoting New and Renewable Energy Solutions in Islands," "Coevolution and Biocultural Diversity Under Globalization," and "Psycho-social Aspects of the Women from Cape Verde Islands Migration in Italy."

Since it is interdisciplinary, my book resembles such conference proceedings, except that it works with a more specific common theme — the negative and positive values of isolation — that ties its chapters together. Some of the useful recent literary compilations on islands include *The Classical World and the Mediterranean* (ed. Giuseppe Serpillo and Donatella Badin; Tema, 1996), *Message in a Bottle: The Literature of Small Islands* (ed. Laurie Brinklow, Frank

Ledwell, and Jane Ledwell; Institute of Island Studies, 2000), and *Beyond the Floating Islands* (ed. Stephanos Stephanides and Susan Bassnett; University of Bologna, 2002). Perhaps my book occupies a space between these multiple-author literary anthologies and Stephen A. Royle's *A Geography of Small Islands: Small Island Insularity* (Routledge, 2001), in that Royle covers many island subjects, linking them by the single theme of insularity. His table of contents includes chapters on literature, geology, history, sociology, communications, politics, economics, and tourism.

More than most of these writers, I am trying to bridge the gap between the humanities and science. These scholarly categories depend upon each other more than their respective representatives may care to admit. Deductive reasoning is fundamental to all fields. Sober, patient, consistent, and long-term inquiry is essential to scholarship, whatever the subject matter. Equally important, in my view, is inspiration — the flashing kind of intelligence that makes connections across borders, whether these borders are around subjects, historical periods, geographical place, gender, race, class, or nation. English Studies is one field that is beginning to recognize the value of interdisciplinary approaches, as seen in the rise of hybrid programs like Cultural Studies and Gender Studies. On a grander scale, universities are beginning to bring back non-specialized degrees such as General Studies and Humanities.

In an era in which interdisciplinary studies are reviving, such spanning of the arts and sciences as my book purports to accomplish suggests one way that such an ambitious endeavor can be approached. Modernism was a difficult master for the writers of its epoch, challenging its adherents to mix new values with old (characteristic of Modernists is their nostalgia for the past). It is fascinating to see how modernist writers' concerns were reflected in, and interdependent on, the problems of their science and social science counterparts. Mead would offer the *fa'a Samoa* (Samoan way) as a microcosmic model to address American family dysfunction, but her projections upon that culture were what contemporary literary theory might call "overdetermined," laden with her own cultural values to the point of unreliability. Darwin's work gave scientists zeal for the scientific method, that might in some sense replace or parallel their faith in a divine creator. The Modernist period was rich with the sense of a quest, a mission, or project to change the world. Whereas one would expect such idealism in fields like literature, or even painting, it is interesting to see the degree of crossover to the sciences during this period. As a bridge between the humanities and science, islands are less arbitrary than they may at first appear. They offer excellent opportunities for description and experimentation (activities of the sciences) and equally significant ground for symbolization, projection, and narration (activities of the humanities). Islands are the place *par excellence* of dreams, so universal a symbol that Melville wrote, "in the soul of man there lies one insular Tahiti," and Dea Birkett echoed him, a century and a half later, with the words, "We all hold a place within our hearts—a perfect place—which is in the shape of an island."

I

Introduction: The Lure of Isolation

Modernism and Insularity

Modernism denotes a period from 1890 to 1930 and describes an atmosphere of skepticism regarding the efficacy of social institutions, such as nation, empire, marriage, church, and school. It also questions the authority and validity of class, gender, and race distinctions. This skepticism extends to literary and artistic forms as well. Victorian novels of social realism give way to the modernist novel that explores the nature of consciousness.[1] Fixed forms in poetry give way to free verse. Classic conventions of drama, such as the Aristotelean unities, are dropped for greater freedom of expression. In art, everyday scenes and objects replace classical subject matter, while form and style are predicated on feeling rather than so-called realism.

My study examines five modernists in five different disciplines—biology, painting, drama, fiction, and anthropology—who shared a fascination with islands. Charles Darwin developed his theory of evolution by natural selection from observation of the biota of the Galapagos Islands.[2] Paul Gauguin forged an individual style of Post-Impressionism during his self-exile in Tahiti. John Synge ignited drama by incorporating the language and experience of Aran Island peasants into his plays. D. H. Lawrence deplored individual isolation but wanted to form an island community of like-minded people located far from the evils of modernity. Margaret Mead took the social organization and sexual mores of Samoan Islanders as paradigms for her own construction of non-traditional families. All five subjects are individualistic, iconoclastic, idealistic, and controversial, both in their own time and in the present. These character traits produce a quality of insularity in their lives which links to island studies in ways that the common etymology of the words *island* and *isolate* suggests. The universal association of islands with the concept of isolation serves as a metaphor for the five subjects of this study. Although in retrospect Modernism appears congenial to revolutionary ideas, no innovator in any time works without opposition. Indeed, opposition often strengthens individuality. These five innovators often worked in isolation motivated in part by cultural resistance to them, and in part by their own necessity to step outside safe parameters in order to make new forms and theories.

The field of Modernism supplies key terms for this study, such as *alienation* and *skepticism*. Postmodernism and its cousin, postcolonialism, supply another set of terms, *the center* and *the periphery*. The periphery is an ideological as well as literal zone, which is inhabited by a disenfranchised and marginalized Other. It is the site of interest and contest for postcolonial studies. Islands are by definition on the periphery of modernity's center, the metropolis. They appeal strongly to both modernist and postmodernist sensibilities. Modernism considers industrial, capitalist civilization inadequate to the needs of artistic or creative individuals. The modern artist is alienated (from Latin *alienus*, other) and needs to protect him or herself from the contamination of industrialized, and therefore mechanized and convention-ridden, society. Modernists found several creative modes of escape or insulation, besides the drug and alcohol abuse which are so often an artist's downfall. More than their predecessors, modernist artists tended to adopt alternative sexual lifestyles, to adventure abroad, and to exile themselves from their homelands. Some combination of these escape routes shaped the life maps of James Joyce, E. M. Forster, D. H. Lawrence, Virginia Woolf, Ernest Hemingway, Joseph Conrad, Henry James, T. S. Eliot, and Ezra Pound, to name but a few English and American high modernists.

Islands as Utopia and Other Fantasies

Island metaphors aptly describe these individualistic behaviors and life choices; in fact, *to island* is a transitive verb meaning *to make something (or someone) like an island; to isolate.* The words *isolate* and *insular* derive from the Latin word for island, *insula.* Islands were a particularly frequent destination of the alienated modernist, but the *pathetic fallacy*, which is the literary term for projecting one's state of mind onto a geographical setting, does not completely account for islands' attractiveness. In other words, seeing oneself as isolated or different does not necessitate living on an island in order to protect that self-identity. Islands exert a unique appeal because they are contained spaces, and therefore, theoretically at least, controllable. Shakespeare's John of Gaunt thought his blessed isle was safe from contamination: "This fortress built by Nature for herself / Against infection and the hand of war" (II.i.45–6). In science, the concept of containment is more concrete: the Galapagos served as a laboratory, or controlled space, in which Darwin discovered a process called *speciation.* In social science, the idea of island as controllable space is more problematic. The human psyche projects its desires as well as its hostility on islands, and finds that it cannot control its environment in the way that a scientist controls his or hers. The smallness of an island tricks the human mind with the hope of gaining full understanding or control of it, but the Other, being a construction of the unconscious, is essentially unknowable.

Another powerful construction of the human psyche is *utopia*, an imaginary place of perfection. From at least the time of Thomas More's *Utopia* (1516), islands have been the key locations for this idealized community, since they appear to be removed from corrupt society. Yet humans do carry the seeds of corruption wherever they go. In *Gulliver's Travels* (1726), Jonathan Swift makes fun of the utopian urge via his gullible hero who visits islands where he finds dystopian communities. The problem is that the community that inspires him, the land of the Houyhnhnms, is administered by horses. Gulliver doesn't fully understand the ramification of this—that humans cannot make utopia because they are human, and therefore fallible. Swift was painfully aware of the human impossibility of creating a utopia, which by its very name (*u-* meaning "no," and *topos* meaning "place") denotes its non-existence.

Besides utopias, islands also figure as testing grounds, especially in children's stories or religious legends. In a French Grail story, *L'Estoire del Saint Graal* (1230), the island is full of terrors (in modern terms, the fears and drives of the unconscious). In his fight against Evil, the hero is tested by God, found to be one of the elect, and entitled to the Grail. *Robinson Crusoe* (1719) is the most famous story of a castaway, its hero an icon of the human condition (of isolation). Crusoe passes his test with high marks, but island tests may also turn out negatively. In William Golding's *The Lord of the Flies* (1954), a small group of English public school boys with good manners turn wild and murderous when stranded on an island without adults. The recent film *Castaway* (2000) parodies *Robinson Crusoe*'s theme of isolation as transformative experience. Instead of a man Friday to order around, Tom Hanks has only a deflated volleyball, which he talks to like a friend. Hanks' amazing transformation from overweight geek to rugged hero, capable of surviving with no more than an ice skate and a volleyball, mocks the kind of totalistic make-over that has become archetypal for the castaway experience. Reality television shows bank on the continuing power of the archetype, pretending that even the losers of *Survivor* are immeasurably enriched by the island experience. The fact that such "isolation" is being filmed by a camera crew ready to provide food, first aid, and transportation in the case of an emergency makes this particular wilderness experience more of a farce than a challenge. In the era of postmodernism, the idea of the castaway experience as necessarily transformative can appear ironic, satiric, and sometimes even ludicrous.

Besides the fantasies of utopia and personal transformation, another compelling attribute of islands is their conquerability. All islands of the world are or have been colonized or dependent on an outside government (Royle 146). The usual small-island political profile is one of "powerlessness, dependency and insignificance" (Royle 134). Islands are extremely vulnerable to attack; once island defenders lose control of the sea, they have to fall back on their own resources and defenses, which are usually insufficient to match the onslaught of larger powers. Shakespeare's *The Tempest* (1611) is literary critics' favorite object for postcolonial criticism since it features a shipwrecked Italian nobleman (Prospero) who steals the land and freedom of an island native (Caliban), whom he compares to a beast. Postcolonial critics read Prospero's actions as a metaphor for British imperialism.[3] A modernist counterpart to *The Tempest* as allegory for imperialism is D. H. Lawrence's "The Man Who Loved Islands" (1927). Its protagonist attempts to control space and human beings. When he finds this impossible, he chooses to die. Whereas Prospero has the opportunity for redemption, Lawrence's protagonist only descends further into psychosis, which leads to his death. Modernism rarely makes islands the unadulterated paradises of earlier literature.

Last, but certainly not least, islands receive a fair share of sexual fantasy projection. Tropical islands are used as sexual lures in the travel industry. They are billed as the perfect place to rejuvenate a marriage or to find new love. Finding love with another vacationer of one's own class and culture is one thing, but finding an exotic Other is often considered more exciting. South Seas, Mediterranean, and Caribbean islands tempt people with the notion of an illicit or different kind of love affair.

Our psychic association of islands with seduction begins at least as early as the eighth century B.C., with Homer's *Odyssey*. Fighting monsters, men, and Poseidon's will took time and energy, but only a fraction of the ten years that Odysseus spent in transit after the Trojan War was actually devoted to these endeavors. The rest of the time he spent with women, or goddesses, the eternal female principle. Two island seductresses were nearly fatal to his marriage. Calypso sequestered him for seven years, Circe, for one year. On yet another island, the Sirens would have claimed the Ithacan crew had they not blocked their ears. Twenty-six centuries later, the fantasy of sexually available, exotic women played a strong role in the decision of Paul Gauguin to live in Tahiti. Nor has the sexual allure of islands lessened in our day. Terry McMillan's best-selling novel *How Stella Got Her Groove Back* (1996) features a middle-aged, female financier from San Francisco who falls in love with a twenty-one-year-old Jamaican. Their affair opens her eyes to the sterility of her yuppie life. It is noteworthy, however, that the couple subsequently makes their home together in California rather than problem-ridden Jamaica. Many of the small islands of the Caribbean are experiencing racial hostility and political instability, making tourism unsafe outside of walled resorts (an island within an island). But dreams die hard, and care-worn mainlanders continue to cherish the ideal of a friendly, erotic welcome.

Island as Metaphor for Self

While islands are metaphors and sites of fantasy projections for utopia, transformation, colonization, and sex, they are also an apt metaphor for the Self. In literature, the island is a powerful icon of the self. From the folk wisdom of John Donne's "No man is an island" to the pathos of Matthew Arnold's "In the sea of life enisled.... We mortal millions live *alone,*" the poetic imagination sees the likeness of human and geographic isolation. For John Donne, people are inextricably connected by their common humanity. He illustrates this connection through the sadness we feel upon hearing a death bell:

> Who bends not his ear to any bell which upon any occasion rings? But who can remove it from that bell which is passing a piece of himself out of this world? No man is an island, entire of itself; every man is a piece of the continent, a part of the main. If a clod be washed away by the sea, Europe is the less, as well as if a promontory were, as well as if a manor of thy friend's or of thine own were: any man's death diminishes me, because I am involved in mankind, and therefore never send to know for whom the bell tolls; it tolls for thee [*Devotions on Emergent Occasions*].

Donne's image of man-as-island works by negation, by calling up an imaginative vision of what man is not. Due to their boundedness, their desirability, and their remoteness, islands constitute a highly imaginable type of space. It is easier to imagine oneself an island than a continent. Donne is well aware that we spend a lot more time imagining ourselves as islands than as parts of a continent. Like his poetry, this sermon operates on an intellectual level, suggesting that while we are not islands in a spiritual sense, we often feel as though we are in our day-to-day existence. The religious feeling of connectedness (what Freud called the oceanic feeling) is so fleeting and so ideal for most people, that its expression in the religious context of a sermon is appropriate.

On the other hand, Matthew Arnold expresses the reverse view — we *are* islands — in the milieu appropriate for that feeling, which is poetry. Despite his early dates (1822–1888), Arnold sounds very like a modernist in his 1852 poem, "To Marguerite — Continued." Characteristically modern is the poignant combination of longing and skepticism (it is a cruel God who creates longing only to frustrate it). In the poem, the speaker laments the isolation he feels when his lover leaves him. He takes his condition to be universal ("*we* mortal millions live alone"). God separated both islands and individuals.

To Marguerite — Continued

Yes! In the sea of life enisled,
With echoing straits between us thrown,
Dotting the shoreless watery wild,
We mortal millions live *alone.*
The islands feel the enclasping flow,
And then their endless bounds they know.

But when the moon their hollows lights,
And they are swept by balms of spring,
And in their glens, on starry nights,
The nightingales divinely sing;
And lovely notes, from shore to shore,
Across the sounds and channels pour —

Oh! Then a longing like despair
Is to their farthest caverns sent;
For surely once, they feel, we were
Parts of a single continent!
Now round us spreads the watery plain —
Oh might our marges meet again!

Who ordered, that their longing's fire
Would be, as soon as kindled, cooled?

Who renders vain their deep desire?—
A God, A God their severance ruled!
And bade betwixt their shores to be
The unplumbed, salt, estranging sea.

In this famous poem, Arnold personifies islands as individuals who once were part of a couple. Aristotle once explained the human sex drive as the compelling urge to link up to another to regain one's original physical status. According to legend, in the beginning, men and women were connected as one physical being. After being sundered, each desperately longed for the other half, and tried to reattach to it via sex. Although Arnold's metaphor of people as islands is superficially the same as Donne's, the tone of his poem is utterly different. For Arnold, our longing for one another is "like despair," because God will always separate us. (Arnold is in a particularly low ebb here, whereas in the companion poem, "Isolation," he asserts that some men believe they can connect with others.) For Donne, human beings are already connected; they have but to remember it when the bell tolls. Unlike Arnold's, the estranging sea plays no overt part in Donne's picture.

Modern and Postmodern Islands

Modernism's tendency to make islands into metaphors of romantic individualism is problematic in post-colonial studies. Since every island is or once was occupied by an outside power, post-colonial studies find in islands a natural topic. Post-colonialism challenges traditional views of islands as inferior (as in Arnold's poem) and conquerable (as in *The Tempest*), or even as isolated. Colbert Nepaulsingh and Derek Walcott describe islands as superior examples of collectivity and multi-culturalism. Fiona Polack describes isolation as a state of mind imposed by the center, and "mirrored back" to the periphery.

In "Writing and Rewriting the Island," Fiona Polack recommends that we consider islands as "havens for the collective"(229). She claims that the characteristic isolation felt by islanders is not due to geography alone but to the projections of the metropolitan center upon them. Like other postcolonialists, Polack reverses the hierarchy of island/mainland or periphery/center binaries, so that the island can become a positive "center" for its inhabitants.

The work of Tasmanian novelist Amanda Lohrey provides another example of the shifting ideology of islands. In her novel, *The Morality of Gentlemen*, she deconstructs conservative tropes of islands by taking away all recognizable indices of islandness. Colbert Nepaulsingh's intervention is similar. In "Islands and Continents," he strikes back at Arnold for privileging continents over islands, using Derek Walcott's work as evidence. In *Another Life*, Walcott likens islands to poetry. Just as poetry has long held a special place among the literary arts, he reasons, so too is island life unique and lyrical. Nepaulsingh even believes that islands are superior to continents, because limited resources on islands necessitate the development of resourcefulness in their human inhabitants.

In "Personal Islands," Syd Harrex reconsiders the island adventure books he read as a child. They were, he now realizes, full of racist and imperialist propaganda. The view of islands they offered was one of isolated places ripe for imperialist conquest. This contrasts with his adult view of the Pacific islands as regional communities. South Sea Islanders are now trying to redefine the image of their home, speaking of "a sea of islands," rather than "islands in a far sea" (56). Quoting Epeli Hau'ofa, author of *The Contemporary Pacific*, he asserts that Oceanic peoples have always conceived of their world as large, not small. Their myths took place in a wide geographical space consisting of sea, sky, islands, and underworld. He urges us to substitute this holistic, Oceanic perspective for the colonial, Orientalist one (the term *Orientalist* was popularized by Edward Said; it means a Western style of dominating, restructuring, and having authority over the Orient). The colonial perspective erects imaginary boundaries, separates interdependent island cultures, and confines people to small spaces. The Oceanic perspective softens boundaries, unites cultures, and broadens the island world.

Five Case Studies

Paradox and tension between the positive and negative values we place on islands are the central themes of this study. At the extreme, we find historical and literary projections of utopia onto islands. Our expectations of island life illustrate our competing desires for togetherness and isolation. On the one hand, islands seem like places where we can make community work, because of their small size. On the other hand, the lack of variety frustrates another human desire. When writer Robert Welch took a vacation on Tory Island, off the coast of Northern Ireland, he saw the potential of island community actualized in that small place. Welch viewed the islanders' courtesy and flexibility as expressions of an environmental necessity:

> We tend to see on islands, society in its most human aspect; because by necessity a small community must evolve ways of coping with individual differences and absorbing them. This is manifest in the readiness amongst island communities to adapt to circumstances. Island communities are less governed by clock-time, Greenwich mean-time; more by tides and their necessitous requirements. But the human community of the island is one which is ready to change arrangements to accommodate variety [71].

For all the optimism of Welch's assertions, the spirit of togetherness that happens on an island is balanced by its opposite — that inexorable sense of claustrophobia, loss, or decay experienced by those who may not literally be stranded, but who feel isolated on an island.

I have chosen my subjects across the various disciplines of science, painting, drama, fiction, and anthropology because I have seen similarities in their concepts of islands and what islands could do for them and for their fields. Despite the isolation inherent in island life, the people in this study shared a longing for human community, and a desire to shape their disciplinary communities, both of which are common drives in a modernist sensibility. As iconoclasts and idealists, each of these writers needed both metaphorical and physical isolation to forge new paths in their respective fields. As a biologist,

Charles Darwin may seem the exception to the rule of the community-seeker, but his famous twenty-three-year delay in publishing the materials collected on the voyage of *Beagle* is an indication of how much he cared about his community. This delay was motivated in part by his distaste for upsetting a community of fellow biologists and eminent Victorians. When he finally did reveal his ideas, he did so in the spirit of contribution to the greater scientific community.

Chapter II investigates the role that the Galapagos Islands played in Darwin's theory of evolution by natural selection. Fellow biologist Alfred Wallace — who had arrived at the theory of natural selection on his own — explains why islands are ideal for studying this phenomenon:

> Islands possess many advantages for the study of the laws and phenomena of distribution. As compared with continents they have a restricted area and definite boundaries, and in most cases their biological and geographical boundaries coincide. The number of species and of genera they contain is always much smaller than in the case of continents, and their peculiar species and groups are usually well defined and strictly limited in range. Again their relations with other lands are often direct and simple and even when more complex are far easier to comprehend than those of continents; and they exhibit besides certain influences on the forms of life and certain peculiarities in their distribution which continents do not present, and whose study offers many points of interest [*Island Life* 241–2].

In an interesting link between biological and human communities, Darwin's travel diary suggests similarities between biological and imperialistic colonization. Strong caution is advised, however, in translating the theory of evolution to the social realm. Social Darwinism — the idea that the races and classes of man evolve from a lesser to a higher state as plants and animals do—flourished at the turn of the century, and was one of the ills that cultural anthropology set out to eradicate. On the other hand, human community at its best is an entity of interdependence somewhat like the ecosystems of the natural world.

Chapter III explores Paul Gauguin's artistic development on Tahiti. During his island exile, Gauguin suffered isolation from the Paris art world, from fellow painters, and from western culture. Yet also on that island he broke through the inhibiting strictures of tradition, even though we are speaking of the already avant-garde tradition of post-impressionism. Through reading biographies, an autobiography (*Noa Noa*, 1901), and paintings, I assess the contributions of this classic tropical paradise towards Gauguin's emancipation. This iconoclast's utopia needed to be far from the metropolis to release him from restricting norms. It also needed to be tropical to allow Gauguin to develop his color schemes and primitive motifs.

As for any traveler, Gauguin's embrace of the Other stemmed from his own loneliness and longing for a type of society that France did not offer. Within a week of his arrival, Gauguin was despondent about Tahiti's chances to revive him; his dream was "brutally disappointed by the actuality" (*Noa Noa* 2). But he did not give up. Both his writing and his painting perpetuate myths about Tahitian Otherness while they simultaneously lament colonialism's ill effects and repudiate his own allegiance with colonial power. Whatever he did not understand about Tahiti, he would call *sublime* or *mysterious*. Although this accorded with the romantic views of his time, this is a prime example of what Edward Said meant by Orientalizing and exoticizing the Other. Gauguin's mystification of Tahiti did little to explain the culture to his readers, but much to enhance his self-myth as artistic and human pioneer in the deep realms of the subconscious. He mystified his role in art as he mystified the foreign culture.

Like Gauguin, John Millington Synge found his signature style only by encountering an island. Aran, the collective name for three small islands in Galway Bay off the coast of west Ireland, helped form him as an artist. In Chapter IV, I examine how Synge's experience gave birth to a new style of drama which defied classic generic divisions between tragedy and comedy. *The Playboy of the Western World* (1907) is his best-known play, partly because of the scandal it produced, and partly because it achieves a difficult balance among various elements of the human condition. Mixing elements and tonalities of both comedy and tragedy, it is truly a tragicomedy.

Central to Synge's literary success is his adaptation of Gaelic—

the language spoken on Aran — into English. It is not difficult to translate verbatim from one language into a related one, but to create believable drama in which characters speak this way takes skill. At the time, it was a touch of genius, because Ireland was seeking ways to recapture its own ethnic identity. Although not quite a speech-island in the linguist's sense, Aran retained Gaelic language and culture far longer than mainland communities. Its oral tradition preserved its history, myth, and legends. The plots of most of Synge's plays came from these legendary tales. The rhythms of his dialogue, the personalities and expressions of his characters, originated here. This culture was a living remnant of the past, but for various political reasons, its dramatization in the Irish National Theater production of *Playboy* embarrassed many audience members. Synge was an innovator whose truthfulness about certain aspects of the human condition (violence, tragedy, and folly) made him unpopular with both staunch Ascendancy Protestants and Catholic Nationalists.

Like Synge's dramas, Lawrence's short story, "The Man Who Loved Islands" (1927), is radical in its portrayal of human extremity. The protagonist's megalomania is already apparent in the opening lines, when the narrator informs us that, "He wanted an island all of his own: not necessarily to be alone on it, but to make it a world of his own" (151). Lawrence's ubiquitous theme throughout his opus— that the will to power is fatal — gains no greater expression than in this one short story. Donne's aphorism, "No man is an island," makes the perfect moral to this story.

Chapter V compares Lawrence's life-long search for good community to the moral precepts taught by the negative example of "The Man Who Loved Islands." Lawrence sought a balance between solitude and intimacy, between idealism and realism, between Rananim (his personal utopia) and the life he actually did lead. Although the short story parodies his own quest for island perfection, he did, throughout his entire lifetime, envision a small, working community that would come together, either on an island or in some faraway land that was not England. His actual living arrangements were a far cry from these dreams, but they reflect his negotiations between shifting ideals and real limitations. Lawrence was more aware than most of the powerful drives of the unconscious, and his island scheme was often consciously a game, a way to let off steam.

Margaret Mead's life and work also illustrate the island paradox of desirable and undesirable isolation. She desperately needed community, and even created a network that she called a world family, yet she began her career by studying remote South Sea Islands. Mead can be seen as isolated in many senses: she was the first woman to make a name in her profession, she went through three husbands, and she plied her trade in the most isolated areas of the earth. Later, she was singled out for possible errors in her Samoan fieldwork; New Zealand anthropologist Derek Freeman made a career out of debunking her. Chapter VI examines Mead's psychological makeup as a reaction against the explosiveness of the nuclear family, her own and others. Her mother was a disappointed feminist whose careers in sociology and public reform were smothered by the demands of a large family and a thoughtless husband. Mead always worked on a large scale. She avoided becoming like her mother, but she did become "Mother of the World," and the most popular anthropologist of all time. The price of her fame, however, was the sacrifice of intimacy and connection with those closest to her.

During the late nineteenth and early twentieth centuries, a modernist ethics of social change and individual freedom permeated fields as diverse as literature and science, and temperaments as diverse as those of D. H. Lawrence and Margaret Mead. Islands function particularly forcefully in the modernist imagination because of the movement's double gaze. Modernists looked forward hopefully to new kinds of community; they looked backward nostalgically to a preindustrial lifestyle that some islands may have preserved. The concept of island serves both visions well. An island is potentially small enough to recreate in one's ideal image. It is potentially remote enough to be safe from modernity's incursions. Neither vision, however, can withstand the test of actual travel; no island can sustain the burden of idealization. The concept of islands frequently detaches from the reality of islands, becoming an imagined elsewhere in human dreams. Such islands are to be found upon an imaginative map that is always in the making. The quest for perfection figured as islands illustrates the desire to create imaginable space for truth and beauty.

II

"Isolated Countries": Darwin and the Galapagos

How Immigrant Species Colonize Islands

The myth that Charles Darwin was a cynic or a misanthrope, or as his biographer Janet Browne puts it, "an evil genius," is curious (1:ix). Reading just a few pages of his diaries, correspondence, travel writing, or even theory dispels such stereotypes. In fact, one may be pleasantly surprised to hear the fresh-faced wonder — at biological, geological, and human phenomena — permeating his account of the voyage of the *Beagle*, undertaken at age twenty-two. Venturing further, one may be relieved to find that even *The Origin of Species* (1859), published when he was fifty, is eminently accessible to the non-scientist reader. The field of evolutionary biology uses the term *selection pressures* to describe forces that cause adaptation and extinction in species. Perhaps his own accessible and reasonable expression was the result of selection pressures existing in the pious Victorian world in which he held the prominent role of respected naturalist and theorist. Previous scientists such as Erasmus Darwin (his grandfather) and Jean Baptiste de Lamarck had already hypothesized that all species are descended from other species, without necessarily eschewing their religious beliefs. But Darwin aimed to generate an accurate description, documented by evidence, of how and why adaptation occurs.

Thomas Malthus's *An Essay on the Principle of Population* (1798), although it addressed human populations, gave Darwin an understanding of competition in the plant and animal worlds he studied. Although he did not understand the genetic mechanism for adaptation (genetics was not yet born as a science), he did produce an extremely accurate description of adaptation, with an abundant number of examples. His later book, *The Descent of Man* (1871), applies this description to humans, thereby stating clearly what had earlier been implied: if man is descended from ape, then God did not create him in His own image in a day or moment.

How did Darwin achieve such a revolutionary feat with such equanimity? Darwin made the unpalatable idea of evolution understandable to fellow scientists by a combination of actions and traits. He was diplomatic towards his critics and fellow scientists. He was thorough in his preparation and collection (including five years on the *Beagle*, and twenty-three years of homework before *Origin* came out). He was strategic in his cultivation of allies. He was characteristically Victorian in his dedication to truth-finding and his belief in the general progress of mankind. For all his efforts, he still lost a few friends (Richard Owen and Robert FitzRoy), but he gained others, and even strengthened his friendships with Joseph Hooker (botanist), Charles Lyell (geologist), and Thomas Huxley (naturalist), who earned the nickname "Darwin's bulldog" for fighting Bishop Wilberforce on the evolution issue. These men defended Darwin and his ideas. He was an established member of their society of elite, London scientists; they respected and liked him.

The genesis of Darwin's theory illustrates the efficacy of the scientific method. Working from small to large—from data to generalization—Darwin eventually found reasons for the changes he observed in species.[1] Although the Galapagos Islands provided a small amount of evidence in relation to the entire number of species Darwin studied throughout a lifetime, his experience there illustrated natural selection in undeniable clarity. The study of island *biota* (the animal and plant life of a region) was the indispensable condition for Darwin's conceptualization of evolution by natural selection because it was through observing island-to-island differentiation of tortoises and finches that he realized that species adapt to their different environments, however slight their differences. He found, too, that the biota of islands considerably distant from the mainland (in this case, about 500 miles) resembles the mainland's more than the nearby islands', even though these are climatically more similar. This proves that biota are transported a very long way; it also disproves creationism and suggests the cause for adaptation, which is the need of intruders (colonizers) to find a niche in existing ecosystems.

What is the theory of natural selection? First, that all species produce many more offspring than can survive on the available resources. Second, life is a competition with other species, in which only the fittest survive. Third, species adapt to their environments in order to survive. Fourth, the process of adaptation is gradual; generation by generation, those characteristics which ensure success are favored. (This contradicts Lamarck's view that creatures possess a will to change, and that they can inherit habits. Mendel later disproved the inheritance-of-acquired-habits theory.) Those individuals and populations with less-favorable characteristics will reproduce fewer offspring, and die or be killed earlier, until they become rare, and possibly extinct. Therefore, each species is adapting, over much time, for the best possible fit it can make to its environment. There is no magical change in the DNA (an element Darwin did not name or know about) in a generation or two; instead, there is the gradual favoring of an already-existing characteristic that some of the individuals of a species possess.

The classic example is the peppered moth, which, during the Industrial Revolution in England, evolved to having black wings, according to a phenomenon called industrial melanism. Its original silvery, mottled wing color was well adapted for camouflage on lichen-covered tree trunks. However, when trees lost this lichen covering due to the soot of industrial pollution, the silver moth was too visible to its predators. A black-winged variety gradually spread and predominated, until it almost replaced the silvery one. Once pollution controls were instituted, the silver lichen returned, and the silver variety of moth made a gradual comeback.

Modern molecular genetics builds on Darwin's base. Although it was available to him at the time of rewriting *Origin*, Darwin apparently did not read Gregor Mendel's 1866 paper on plant hybridism, or else he went along with the common misconception that Mendel's experiment with pea plants was an anomaly (Leakey 23). Mendel theorized the existence of dominant and recessive genes, though they were not to be called genes until 1909 by Wilhelm Johannsen. Mendel's fundamental contribution was distinguishing phenotypes from genotypes. He recognized that there are both visible characteristics (phenotypes) in the plant or animal, such as eye color or height, and invisible elements in the gene itself (genotypes). A gene may be recessive in a given generation; therefore its phenotype is not present. It takes two recessive genotypes in offspring for the recessive trait, or phenotype, to reappear. Mendel's major correction to Lamarck and Darwin's thinking was that parental traits do not blend in offspring; they are unequally expressed, because of the phenomenon of recessive genes.

When DNA was discovered in 1953, Mendel's theory was expanded into what Richard Leakey calls the Central Dogma of DNA: "somatic (bodily) events cannot be translated into changes in the individual's genetic information" (19). In simple terms, this means that habits have to be passed on by parental teaching and acquired experience; they cannot be programmed in the DNA of an individual or species. After DNA, Lamarckism was "dead, and [would] never be resurrected" (Leakey 28). Darwin's speculation of "use-and-disuse heredity" was wrong, but he was right in noting that no matter the mechanism, natural selection did occur.

Besides its chapter on natural selection, *The Origin of Species*' chapters on geographical distribution are relevant to this study. In them, Darwin recognizes the relative importance of isolation in the creation of new species, thus linking his projects with our interest in islands. The book's wording closely tracks the relevant chapter ("Galapagos Archipelago") in *The Voyage of the Beagle* (first edition 1839), thereby emphasizing the importance of that initial illustration of adaptation. By 1859, Darwin is able to confidently express the Galapagos material as scientific fact. The reason why *speciation* (development of new species) occurs so frequently on islands is that "species occasionally arriving after long intervals in a new and isolated district, and having to compete with new associates, will be eminently liable to modification, and will often produce groups of modified descendants" (*Origin* 380). This is the theoretical expression of his earlier field observation of the gradations in the size and shape of finch beaks on the different Galapagos islands: "Seeing this gradation and diversity of structure in one small, intimately related group of birds, one might really fancy that from an original paucity of birds in this archipelago, one species had been taken and modified for different ends" (*Beagle* 328). The trepidation of the 1839 wording—"one might really fancy"—vanishes in the bold 1859 assertion that "species will be eminently liable to modification."

The other Galapagos phenomenon that impresses him is the similarity of its biota to South America's, a continent with extremely different physical conditions. His amazement is not due to any incomprehension of the means of transport, for he refers to the possibility of floating rafts, composed of tree trunks, brush, and other debris, as carriers of immigrant species, a phenomenon which is by now well-documented. His astonishment is directed at this further proof of the adaptations that continental species undergo once they begin to colonize distant oceanic islands. Galapagos species are related to continental species, but have undergone adaptations to the climate, vegetation, and competition posed by the ecology of the site of colonization. New species adapt to fill a niche in existing ecologies.

In *Voyage of the Beagle*, he asks the question—why is Galapagos biota similar to South America's—which he answers twenty-five years later, in *The Origin of Species*:

> Why, on these small points of land, which within a late geological period must have been covered by the ocean, which are formed of basaltic lava, and therefore differ in geological character from the American continent, and which are placed under a peculiar climate,—why were their aboriginal inhabitants, associated, I may add, in different proportions both in kind and number from those on the continent and therefore acting on each other in a different manner—why were they created on American types of organization? [340].

And *Origin of Species* answers:

> I believe this grand fact can receive no sort of explanation on the ordinary view of independent creation; whereas on the view here

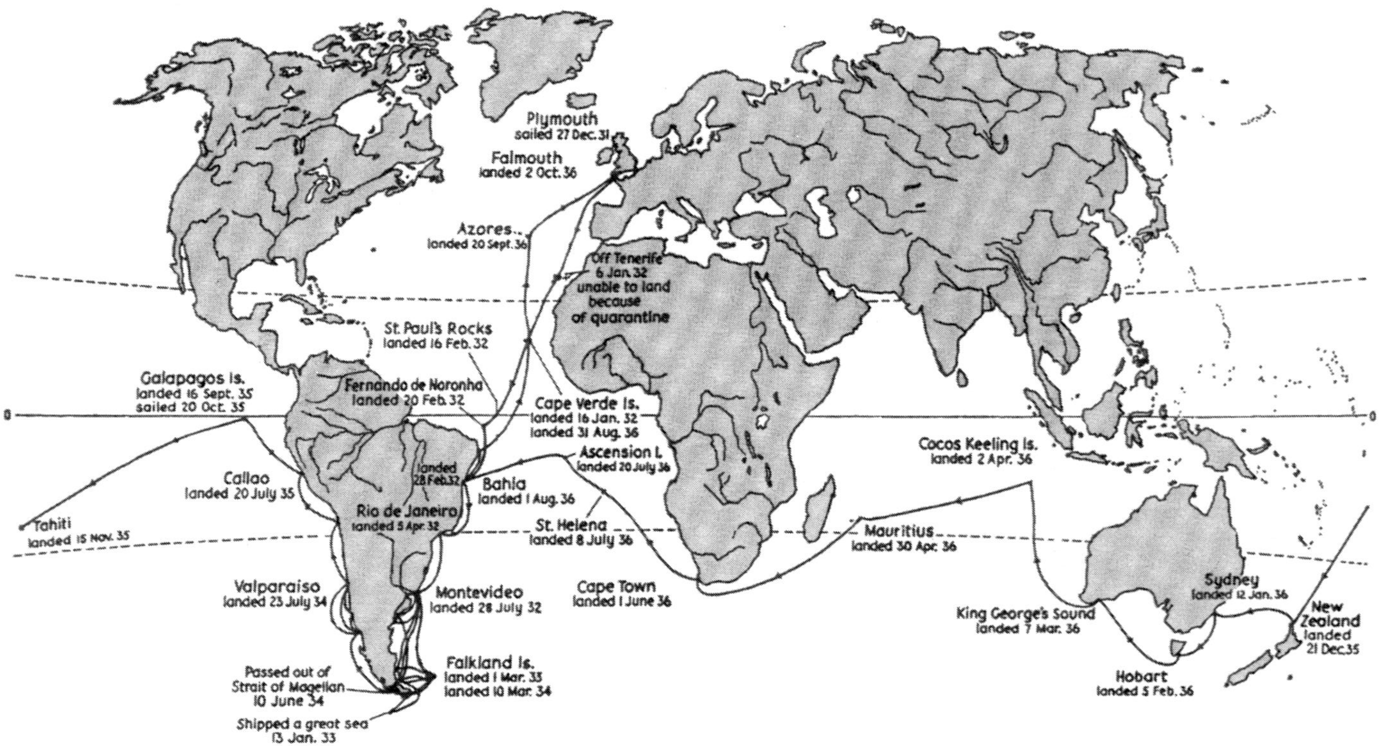

The Voyage of the *Beagle* 1831–1836

> maintained, it is obvious that the Galapagos Islands would be likely to receive colonists, whether by occasional means of transport or by formerly continuous land, from America ... and that such colonists would be liable to modifications;—the principle of inheritance still betraying their original birthplace [386].

Twenty-five years of cogitation, further experiments at home, acceptance in London's scientific societies, and ongoing conversations with the top English and American scientists finally give Darwin the confidence to say what he already knew at the Galapagos, but expressed in terms of wonder rather than science. The end of the voyage finds him puzzled about the distinction between varieties and species, but it is the kind of puzzlement that is natural before a thinker breaks the mold. After five years' observation of different islands and continents, Darwin's mind still returns to the Galapagos as the best illustration of adaptation, as he writes:

> When I recollect, the fact that from the form of the body, shape of scales & general size, the Spaniards can at once pronounce, from which Island any tortoise may have been brought, When I see these islands in sight of each other, & possessed of but a scanty stock of animals, tenanted by these birds, but slightly differing in structure & filling the same place in Nature, I must suspect they are only varieties.... If there is the slightest foundation for these remarks the zoology of Archipelagoes—will be well worth examining; for such facts would undermine the stability of Species [Browne 1: 339].

The mold he was breaking was the creationist belief that species are fixed and eternal groups. Mutability of species severely challenged beliefs about God's perfection and His creation.

By now, the concept of speciation is a commonplace. Also well-accepted are Darwin's dicta about islands, which laid the foundations for the field of island biogeography. This field recognizes that islands have special properties for the study of evolutionary processes.

Opposite: The Voyage of the Beagle 1831–1836. "Voyage of the Beagle: Map, page xvii" from _Darwin_ (NCE), second edition, edited by Philip Appleman. **Copyright 1979, 1970 by W.W. Norton & Company, Inc. Used by permission of W.W. Norton & Company, Inc.**

Barry Cox and Peter Moore define islands' role in understanding evolution:

> Isolation is one of the key factors in permitting evolutionary change, for it allows the gene pool of a population to become different from that of other populations. On large land-masses, that isolation is variable in its nature and therefore in its effects on the biota as a whole. Islands provide clearer examples of isolation, for the sea surrounding them is an environment in which few terrestrial or freshwater organisms can survive for any length of time. Special adaptations for transport by air or water are necessary for an organism to cross a stretch of ocean. Dispersal to islands is therefore by a sweepstakes route, the successful organisms sharing adaptations for crossing the intervening region rather than for living within it. This greatly restricts the diversity of life that is capable of emigrating to an island. But ... many factors control precisely how many organisms will reach and colonize an island, and also control the degree of diversity that its biota will ultimately attain. By comparing the biotas of islands of different size, or islands lying at different distances from their source of colonists, or islands of different topography or lying at different latitudes, we can learn much about the interaction of these factors in the control of organic diversity [105].

Although Darwin's ideas about islands inform this 1985 textbook passage, the purpose of island biogeography is different now, since evolution has already been amply proven. The key to this difference is the word "control" in the last two sentences, even though it is used there ostensibly in the sense of natural control by selection pressures. The chapter from which the quoted passage is selected, "Life (and Death) on Islands," makes an important corrective to the quantitative Theory of Island Biogeography formulated by ecologists Robert MacArthur and Edward Wilson in their 1967 book. The Theory maintains that it is possible to predict the rate of colonization and extinction of species on an island by calculating an "equilibrium number of species" for any given island whose size and degree of isolation are known. These numbers are then utilized by ecologists for interpreting changes in the biota of nature reserves or national parks and for suggesting how these should be managed (Cox and Moore 116). Cox and Moore question the validity of assuming that island biogeography can be applied to isolated sections of mainland, as MacArthur and Wilson do. The quantitative Theory has also been criticized for assuming that the physical characteristics and degree of isolation of an island are the only factors influencing the equilibrium point of species. There are other important factors, such as the biological interactions of species or the possibility of an increase in the number of species by evolution rather than immigration (Cox and Moore 116). Curiously, the 1967 Theory ignores the very things that Darwin stressed — the struggle for existence and the survival of the fittest. Cox and Moore conclude that it may not be feasible to create mathematical models of biota, which are by Darwin's definition (mutability) so infinitely changeable.

The Royal Mission: Another Type of Colonization

Her Majesty's Ship *Beagle* set out to circumnavigate the globe in 1831, with two royal missions: to chart the southern coasts of South America for British commercial and imperial uses, and to return three Tierra del Fuegans to their native land with a missionary who would help them to convert the rest of the tribe. Biology was only a sideline. As we might expect, Captain Fitzroy had no problem with the government's imperialist agenda, but Darwin did register his discomfort about colonial enterprises. His recorded observations of the voyage do not stop at zoological data; his commentary on human populations is every bit as interesting, and often more so. Although his reactions to colonialism are predictable for a person of his class, he did not accept the imperial ethos without question. In fact, he frequently contradicts himself on matters of colonial intrusion. He asserts both the efficacy and the cruelty of British hegemony, without suggesting that it should be otherwise: "All the aborigines have been removed to an island in Bass Straits, so that Van Diemen's Land enjoys the great advantage of being free from a native population. This most cruel step seems to have been quite unavoidable, as the only means of stopping a fearful succession of robberies..." (387). The words *enjoy* and *free* suggests he shares the administrators' intolerance of natives, whereas calling the removal a cruel step means he sympathizes with the natives. But Darwin does not seem particularly concerned about resolving his inconsistencies. He may have toed the political line in order to remain within his patriotic Captain's good graces. Later, he would again have to consider politics when deciding when, whether, and how to reveal the processes of natural selection to a pious world.

Although Darwin found certain colonial abuses distasteful, he adhered to most of the beliefs and values of his class that many people now find distasteful. He believed in the good of missionary work, even though his own captain's project in that arena was unsuccessful. He felt proud of England for establishing civilization in remote places, a civilization he measured by the ratio of farms and clothes to natives (he resented the lack of proper costume on the Tahitian women).[2] On the other hand, he abhorred slavery, and argued with FitzRoy about it. Nor did he agree with the Captain's use of flogging as ship discipline. The Captain and Darwin were vivid personifications of different social classes. Captain FitzRoy was an aristocrat with at least one infamous relative: his uncle, Viscount Robert Stewart Castlereagh, had committed suicide after a troubled political career, fearing that top members of Parliament and society thought he was homosexual. In allegiance to his upbringing, FitzRoy was Tory, Anglican, and a proud Navy man from the age of fourteen. Charles Darwin was from a wealthy, untitled, middle-class family. He was a Whig with abolitionist relatives, and a budding atheist at twenty-two. He formed an academic identity by ingratiating himself to Cambridge society.

In his diary, Darwin remarked on FitzRoy's aristocratic assumptions. The Captain enjoyed a great sense of entitlement. He expected preferment in the Navy, deference from the lower classes, and acceptance of his views by everyone on board ship. Initially, Darwin admired his strength of character, remarking on his "cultured authority and direct way of speaking" (Browne 1: 159). Later, he found that his Master's mercurial temper created problems aboard ship. FitzRoy's famous temper would be directed at Darwin some years later, when he publicly decried evolutionary theory. The relationship between these two adventurers has fascinated more than one reader.

Richard Lee Marks' *Three Men of the Beagle* is a fictional version of the *Beagle* voyage, as told by three of its participants: Captain FitzRoy, Darwin, and Jemmy Button, the Fuegian native on whom FitzRoy pinned his hopes for saving savage souls.

Charles Darwin did not set out to be an evolutionary biologist, nor an iconoclast of any stripe. Born to a Darwin and a Wedgwood, he was the product of successful bourgeois capitalism (Wedgwood pottery was the business concern on his maternal side) and of successful professionalism (his paternal ancestors were reputable physicians). His father was also an investor and a lender, and his grandfather, Erasmus Darwin, was a poet who also wrote a theory of evolution (which we now know to be marred by a misunderstanding of genetics). Since Charles' performance was mediocre in both secondary school and his first two years of medical school at Edinburgh, his father transferred him to Cambridge to study for the ministry. This field did not interest Darwin, but field biology did. His enthusiasm for collecting and his amiability attracted the attention of Professor John Stevens Henslow, who became his mentor. Henslow provided the recommendation for Darwin to work on board the *Beagle*. He must have seen not brilliance (he said Darwin had not achieved that yet), but a capacity for adventure, and the good manners necessary for the hierarchy of a Navy ship. He told Captain Robert FitzRoy that Darwin was the best candidate at Cambridge for the position of ship naturalist. He knew that FitzRoy's chief requirement was a good companion — in other words, a gentleman.

FitzRoy was a conservative; his loyalty to Navy, Crown, and Church went unquestioned. But he held some liberal views. He loved and admired the Indians and other natives whom he met on the voyage, even though he also tried to Christianize them. To this end, he transported Jemmy Button and two other Fuegians to England, converted them, and returned them to Tierra del Fuego, instructing them to teach English and convert the rest of the tribe with the help of a volunteer missionary named Richard Matthews.

The project was not a success. Its main hope, Jemmy Button, was unable to change the values and habits of his people. After a week, the missionary begged to return to the ship, as it appeared that the Fuegians were on the verge of killing him. He was not exaggerating. In 1859, Indians murdered half a dozen missionaries who were trying to establish a church on Navarin Island. In the late 1860s, a white missionary was finally able to live among the Indians without losing his life, but the Indian population was dwindling. By the 1960s, the Yahgan Indians were extinct. Captain FitzRoy had not understood the insurmountable obstacles to cultural imperialism in this frigid outpost, 600 miles north of Antartica. It was inhospitable to agriculture, and its peoples practiced cannibalism.

In his conclusion to *The Voyage of the Beagle*, Darwin identifies the "scenes and objects" of Tierra del Fuego as making the greatest impression out of all the places and people encountered on the five year voyage. He writes of the scenery in terms reminiscent of his beloved John Milton, whose *Paradise Lost* he carried on the journey: "primeval forests undefaced by the hand of man ... where Death and Decay prevail" (436). It was easier for Darwin to imagine the mutation of a dinosaur to an ostrich, say, than a savage to a civilized man: "perhaps nothing is more certain to create astonishment than the first sight in his native haunt of a barbarian — of man in his lowest and most savage state ... and part of the interest in beholding a savage, is the same which would lead every one to desire to see the lion in his desert, the tiger tearing his prey in the jungle, or the rhinoceros wandering over the wild plains of Africa" (437). The vision of naked, painted Fuegians, wildly gesticulating and ululating, running till their noses bled and their mouths foamed, never left him. If the great tortoises of Galapagos fueled his imagination for the idea of natural selection, the wild men of Tierra del Fuego fueled that most iconic of Darwin's drawings, the chart of man's lineage. The ape evolves to the cannibal, who evolves to the civilized man. The best-known books of his later years, *Descent of Man* (1871) and *The Expression of the Emotions in Man and Animals* (1872) came out of that powerful image of Fuegians.

Tierra del Fuego and the Galapagos Archipelago were the most influential landings on Darwin's trip, but other islands made impressions that likewise led to intellectual discovery. Keeling Island, the Falklands, Tahiti, and New Zealand rank highest in this regard. The smallest of these, Keeling Island, proved a theory of coral formation that Darwin had already conceived before seeing a coral island; he admits this is the one time he proceeded deductively instead of inductively (Browne 1: 318). Darwin's favorite geologist was Charles Lyell, the proponent of uniformism in geological studies. Lyell went against his fellow scientists in believing that earth's changes were slow and gradual, though caused by underground geothermic changes. FitzRoy gave Lyell's *Principles of Geology* to Darwin at the beginning of the trip, and Darwin used it to study islands and continents that they visited. He felt vastly indebted to Lyell's theory of uniformism, using the concept of "slow and gradual" change in his own theory of biological evolution. Consequently, it was with all due respect that Darwin modified Lyell's description of coral island formation.

Lyell believed that coral islands were created by the accumulation of coral on top of submerged volcano craters. Coral polyps could only live in the top few fathoms of water, where there are sufficient light and warmth for their survival. Darwin did not believe they had to be founded on top of craters. He applied Lyell's theory of the "self-regulating oscillatory movements of the earth's crust" to the formation of coral islands (Browne 1: 317). Mountain masses are evidence of thrust in a continent; nearby ocean floors subside in compensation for their thrust. To every geological action there is an opposite and equal geological reaction. If the Pacific Ocean were regularly subsiding, then there was no need to imagine volcanic craters as undersea bases for coral reefs; it was sufficient to consider that islands subside as the ocean floor they stand on subsides. Hence, coral reefs grew up on island tops as the islands themselves gradually sank under the sea's surface. He proved this theory by taking soundings to discover the content of the coral reef bases, and published an article which Lyell graciously acclaimed.

This story demonstrates several aspects of Victorian science. In the nineteenth century, there was still undiscovered territory, and thus a greater opportunity for scientists with sufficient interest, determination, and money to travel to field sites (the latter tended to narrow the field to gentlemen). Theoretically, the geology of coral islands could have been discovered by anyone equipped to take soundings. Compared to today, the boundaries between the fields of natural science were more fluid; it was possible for Darwin to train in biology, teach himself geology, and learn math from Captain Fitzroy (in order to read ship chronometers). No one thought it strange that Darwin would research and publish in biology, botany, and geology. In the story of coral islands, we can see how his study of geology enhanced his comprehension of biology and vice versa. Although there is a movement towards academic interdisciplinarity today, it faces an uphill battle against disciplinary territorialism and the high value placed on specialization.

The story of Keeling illustrates small islands' capacity for providing fertile imaginative space for the construction of theory. The smaller a geographical entity, the easier it should be to determine how it is made. This may not be true in all cases, but gives rise to a second principle. The best way to solve large problems is to break them down into smaller problems. Isolation simplifies and gives a clearer mental picture of both abstract and concrete phenomena. Islands attract inquiring minds, like Darwin's, because they paint vivid pictures of problems to be solved.

In contrast to Keeling Island and coral reef formation, the problem of the Falkland Islands was never solved in Darwin's time. They were another island community that raised both biological and political issues for him. The biological question was whether the slightly different foxes found on the East and West Falklands were varieties or species ... a significant question when we consider its relation to speciation. Can a variety morph into a species by adaptation and extinction? The political question was, to whom or what did these islands rightfully belong? Darwin does not deeply engage the political question, but conveys a literary sense of irony:

> After the possession of these miserable islands had been contested by France, Spain, and England, they were left uninhabited. The government of Buenos Ayres then sold them to a private individual, but likewise used them, as old Spain had done before, for a penal settlement. England claimed her right and seized them. The Englishman who was left in charge of the flag was consequently murdered. A British officer was next sent, unsupported by any power: and when we arrived, we found him in charge of a population, of which rather more than half were runaway rebels and murderers [*Beagle* 162].

The Englishman who dies while guarding his country's flag symbolizes the follies of imperialism. Falkland tensions continued to mount after Darwin's visit, some of them caused by his Captain. FitzRoy's missionary successors used Keppel Island (off the coast of West Falkland) as a colony to educate and Christianize Fuegian hostages. The plan was to re-introduce these "improved" savages back into their Fuegian tribes, hoping that they would spread the word of God. Instead, they massacred the whites at the Navarin Island church in 1859, the year of *Origin*'s publication. FitzRoy had been depressed by news of Jemmy's regression to savagery; this larger scale horror must have been even more troubling to him. The vicissitudes culminating in FitzRoy's suicide in 1865 were several, but the rumor that his friend Jemmy instigated the Navarin Island massacre must have been a substantial blow.

Connecting Humanities and Science

Since both the biological and sociological information acquired during his trip impressed, influenced, and changed Darwin, it is tempting to draw a relationship between these traditionally separate halves of academic discourse, scientific and humanistic. *Voyage of the Beagle* gives equal time to each subject, with a noticeable increase in human interest towards the end. But Darwin's tone changes when he discusses sociology or politics. His characteristic scientific detachment does not carry over to his observations of human culture. Darwin was not pretending to be an anthropologist (the field had not yet been formally invented), but one would expect that his scientific attitude might carry over. He did many of the things that anthropologists do. He collected data on fascinatingly different peoples: the Gauchos of the Pampas, the Guasos of Chile, the Fuegians of Tierra del Fuego, the Tahitians, Maoris, and Australians of the South Seas. His imagination was stirred as often by human problems as biological ones. He pondered the subsistence-level existence of the Fuegians and lamented the attempted extermination of South American Indians by white conquistadors. But he failed to apply his understanding of adaptive techniques to human populations, and there was no academic discipline to help him. He was lacking that Central

Dogma of the twentieth century—cultural relativism. He does not acknowledge his cultural prejudices, as we have learned to do today. To contemporary ears, the following statement sounds ironic and ethnocentric: "To hoist the British flag, seems to draw with it as a certain consequence, wealth, prosperity, and civilization" (*Beagle* 438). Ending his first book on this patriotic note, he appears to have forgotten his earlier image that represented to him the absurdity of the colonial project — the Englishman murdered on the Falklands for his association with the British flag.

In a subtle way, Darwin does mesh the social with the biological. One cannot read about the biological "colonization of islands by immigrants" without noticing the shared discourse of postcolonial studies and evolutionary island biology. This is not surprising when we remember that it was initially Malthus's study of human population that supplied the missing link for Darwin in his theory of natural selection. The human "selection pressures" of war, epidemic, accident, and famine control population growth. Translating Malthusian terms to biology terms was easy for Darwin. Predators, food and water supply, climate factors, availability of mates, environmental catastrophes, and pollution are the selection pressures that control population in the Animal Kingdom.

Likewise, the biological words and concepts of *indigeneity* and *colonization* are borrowed from the human realm, and the struggle for survival applies to humans, as well as animals and plants. When he perceives that acts of colonization kill natives, Darwin is talking about the struggle for survival at a human, social level: "Wherever the European has trod, death seems to pursue the aboriginal. It was melancholy at New Zealand to hear the fine energetic natives saying, they know the land was doomed to pass from their children" (qtd. in Beer 132). This statement directly contradicts the one quoted above, which asserts that British colonists consistently spread wealth, health and happiness. Part of Darwin's attractiveness as a subject of study is his double consciousness that results in self-contradiction. Does he believe that wherever British colonists go, they spread wealth, health and happiness? Or does he believe that wherever his compatriots go, they bring death to the indigenous people? It might be a little bit of both, since humans make attempts to adapt (to colonization) rather than dying. Those who adapt to colonization will find a niche in the ecology of the new system. Jemmy Button did; he became FitzRoy's middle man between civilization and primitives. But was he happy? Not if we believe the rumor that he led the massacre. On the other hand, his compliance resulted in food and gifts— no small consideration for one in his hostile living environment.

In her article, "Writing Darwin's Islands," Gillian Beer argues that *Voyage of the Beagle* was as important to the development of Darwin's political consciousness as his biological theory. Witnessing the horrors of colonization (including slavery) made Darwin question England's aristocratic assumptions. Beer points out that the disjunction between what Darwin thinks about England (his received knowledge) and what he sees on his voyage is never resolved: "the disturbance never settles" (134). His rhetoric is either consciously or unconsciously equivocal whenever he describes a scene of colonialism or extirpation (such as General Rosas' campaign to extirpate the Argentine Indians). Trying to have it both ways, "the writing registers a deep unease, a disequilibrated reach for a vocabulary that in the event cannot steady itself along one axis of judgment" (131).

English people have a long tradition of pride in their island nation. In *The History of England* (1913), historian Thomas Macaulay proclaimed that islanders are superior beings, and that England is the best island in the world. According to Beer, Darwin had to recognize this characteristically English sense of superiority before he could mature enough to write *The Origin of Species*. She thinks he does achieve a modicum of cultural relativism, which was helpful in his task of countering the scientific absolutes of his time. Like myself, Beer sees Darwin's literal islands (the Galapagos, Tierra del Fuego, and Falklands) as intricately connected with the various kinds of metaphorical islands that he needed to confront in order to achieve success in what she calls "an imaginative and argumentative revolution" (139). Darwin also constructed his own metaphorical island of self-protection. The second half of his life is curiously similar to that of the other subjects of my study, insofar as he isolated himself in his country laboratory.

My analysis of both biological and social factors intends to clarify islands' roles in Darwin's life. The Galapagos experience was the indispensable condition for Darwin's synthesis of evidence into the theory of natural selection, and the entire voyage of the *Beagle* fed the intellectual and spiritual growth that made possible the writing of *The Origin of Species*. Islands are the main players in the story of this voyage. They are the destination, the laboratory, the field, the raw material, and the imaginative essence. They are of course real, but they are also imaginative space for the discovery of nature's laws. Darwin's voyage and his subsequent theorization of evolution by natural selection illustrate the imaginative power generated by islands.

III

"To Dream Before Nature": Gauguin in Tahiti

> A myriad of gingerbread men cavort in hyperbolic landscapes characterised by a savagery for which no perceptible phenomenon provides an excuse.[1]

As these lines from an 1893 critical review suggest, Gauguin's art has often been severely misunderstood and flamboyantly ridiculed. The word choices of this reviewer, Charles Merki, reveal more about himself and the moral climate of the *fin de siècle* art world than about Gauguin or his Tahiti paintings. The connotations of the words *gingerbread* and *savagery* are racist to today's ear; the failure to acknowledge that most of Gauguin's figures are women, not men, prudish; the word *cavort* implies a desultory glance at the paintings, since the figures are still and solemn. To call the landscapes *hyperbolic* suggests he did not understand the fundamentals of Post Impressionism, which state that the expression of emotion, not mimesis, is the painter's goal. Gauguin expressed this principle by calling art an abstraction: "derive this abstraction from nature while dreaming before it, but think more of creating than of the actual result" (Kelder 155). Gauguin was not aiming for verisimilitude. Towards the end of his life, he even admitted the "enormous mathematical faults" in his masterpiece *Where Are We Going,* yet asserted their necessity, because only when emotions "burst forth like lava on the canvas" can true artistic integrity be achieved (Goldwater 112).

Mr. Merki's sententious criticism succinctly captures the prejudices Gauguin faced. He fought the Paris art world and bourgeois ideals, creating a self-legend that made certain aspects of his life better known than the paintings themselves. For instance, it is well known that he left his wife and five children to fend for themselves, that he took Tahitian *vahines* (women; wives) despite his syphilis, that he led a cult of possibly homosexual young artists in Port Aven. He has also been accused of stealing Synthetist techniques and calling them his own, of aggravating van Gogh's mental illness, and of alienating every friend he ever made. His editor and publicist, Charles Morice, and his dealer and patron, Daniel de Monfried, were the only two who stood by him till the end. The purpose of this chapter is neither to perpetuate nor to disprove gossip about Gauguin's personal life, but to show that Gauguin's particular temperament (artistic, iconoclastic, proud, impulsive) thrived on self-exile from the Paris art world.

Much has been written about Gauguin's life and art; the challenge to any scholar is to find a new perspective. Recently, two studies have examined Gauguin's sexual and colonial attitudes, coming to very different conclusions. Stephen Eisenman's 1997 *Gauguin's Skirt* seeks to balance the hyperbole of the debate on Gauguin's worthiness by introducing third terms into various aspects of the conflict. Eisenman proposes that Gauguin's sexuality was liminal, hovering on the border between genders, and making him receptive to the *mahu* (homosexual or third sex) in Tahitian culture. As an androgyne himself, he appreciated the quality in both men and women, and certainly found much androgyny in Tahiti. This counters feminists' argument that Gauguin misogynistically represented female nudity for pornographic purposes. Eisenman even finds a positive political component to Gauguin's interactions with the Tahitians. He believes that Gauguin "transported to the colonies" Rousseau and Diderot's ideal primitivism, "where it might eventually do some good" (205). Today, in their struggle for independence, Tahitians creatively utilize certain positive aspects of the stereotype of primitivism to assert their own value.

On the other hand, Nancy Mowll Mathews' 2001 biography, *Paul Gauguin: An Erotic Life*, argues that Gauguin's eroticism is tinged with violence. She is determined to prove that Gauguin's abusive nature dictated his life and art,[2] and that his Tahiti paintings "show practices that were often perverse and painful, yet strangely beautiful" (178). As evidence she offers only a painting of a man beating a woman (*Tahitian Love*), and the famous nude, *Manao Tupapau*, which I will discuss at length in a later section.

Although I do not intend to avoid the contention about Gauguin's personality, it is not my first focus. I am more interested in how Tahiti helped Gauguin to achieve his own style. What did primitivism mean to Gauguin? Can one be primitivist without being racist? Why was Gauguin's first Tahitian sojourn (1891–1893) his pinnacle of success, rather than the second (1895–1903), when he was more professionally secure? Modern art is characterized by the formation of small, intense schools in which artists share ideas, and borrow images and techniques from each other, thereby cross-fertilizing new approaches to the field. Why did Gauguin move away from the Synthetist school that he had helped create? How did the islands help him create his own best style? How did his island experience compare to Darwin's?

Isolating Moves

Like most artists, Gauguin was driven by both his rebellion from established values and his intense need for acknowledgement as an artistic genius. His life is a series of steps away from tradition, family, and convention (linking him to Lawrence's "man who loved islands," who moved to successively more remote places). Since they caused him to abandon networks and community, these departures led inevitably to personal isolation.

Gauguin's first decisive break from convention was quitting his job as insurance agent to paint full time in 1883. In 1885 he left his wife and five children. In 1886 he moved to the village of Pont Aven in Brittany to paint the primitive motifs of Breton peasantry. In 1887 he traveled to Martinique with his friend and admirer Charles Laval for a business opportunity that never materialized and an artistic opportunity that did, though fever made him leave the island before realizing it. Back in France, Gauguin moved from Port Aven to the more remote Le Pouldu, to disengage from its tight-knit, yet competitive artistic colony. The attempt to break his isolation by moving in with van Gogh at Arles in 1888 (though there were other motives as well) ended disastrously with van Gogh's self-mutilation. After a fight with Gauguin, van Gogh cut off his ear and sent it to a prostitute. Though Gauguin's rough behavior did not *cause* van Gogh's mental illness, the ear-cutting incident was the first of several episodes which finally ended in his 1890 suicide.

Gauguin's greatest break from convention was his 1891 decision to move to Tahiti, where, he wrote, "I shall devote myself to great endeavours in art, far removed from all artists' envy" (Becker 23). He was also fleeing something more intangible — European *ennui*. He believed he could restart his life in a place free of European vices:

> If only the day would come when I can flee into the forests of a lonely island in the South Sea like a savage, to live there in peace and ecstasy for my art alone. Surrounded by a new family, far away from the European struggle for money. There in Tahiti, I will be able to listen in the silence of the lovely tropical nights to the soft murmuring music of the movements of my heart in loving harmony with the mysterious beings in my entourage. Free at last! Without money problems I shall then be able to love, to sing and to die [Becker 22].

Nearly every component of the typical European white man's myth of tropical paradise is contained in this famous letter to his wife Mette.[3] Tahitians never go hungry because food is free, to be picked from the trees and sea. Tahitians live in harmony, without greed or envy. Tahitians don't care about money. Tahitians will give him a new family. In actual fact, it *was* extremely easy to get a woman, as both young girls and their parents approached Gauguin with offers, but he never did construct a new family, even though he sired four or five more children.[4]

Gauguin's reentry into Paris during the years 1893–5 was a painful experience. He made print illustrations for his book, *Noa Noa*, but did not paint. Society excluded him, and he relished the opportunity to flaunt his unconventionality. He lived with a teenaged

Javanese girl. Their bizarre dress and pet monkey inflamed the tempers of simpler people, such as the stevedores who assaulted him, and broke his leg at the Brittany seashore in 1894. He sold eleven paintings at his one-man exhibition and received an inheritance from his uncle, but still he did not feel secure.[5] Despite the preponderance of negative circumstances, he did achieve one strangely valuable thing during this sojourn, and that was notoriety. To cultivate his bad-boy image, he held weekly salons at his studio, displaying Polynesian art objects, telling titillating stories about his sex life on the island, and boasting that he had "gone native."

Criticisms received during his one-man show at Durand-Ruel in November of 1893 caused Gauguin to make his final break from Paris. According to Charles Morice, this prestigious opening was the moment when detractors finally broke his spirit:

> In the spacious gallery, where the walls were aflame with the visions of his paintings, he stood observing the people, listening to them. Soon he was certain: none of them understood him. That was the final break between him and Paris. All of his great plans had failed, and what was perhaps the hardest blow for this proud man: he was compelled to admit that he had crafted his plans poorly. Had he not seen himself as the prophet who, misunderstood by the philistines who had failed to realise that they should bow before him, takes his leave to find a new perspective, only to return again greater than ever before? [Becker 80].

During the exhibit, Gauguin antagonized Paul Durand-Ruel, the most influential modern art dealer in Paris, losing his opportunity for an important sponsor. Although he would always rationalize and excuse his temper tantrums, Gauguin must have realized they did him irreparable harm. About this time, he began prosecuting his grievances in courts. He sued the Brittany stevedores for assault and his Le Pouldu landlady, Marie Henry, for the return of paintings he had given her. Well before his return to Tahiti in September 1895, Gauguin knew that he must make a final break from France. His second departure for Tahiti was not a heroic journey; it was an exile, necessary for his survival.

The six years Gauguin spent in Tahiti from 1895–1901 were nothing like the first eighteen months. Despite his wish to become Europe's most famous noble savage surrounded by an "entourage of mysterious beings," Gauguin consistently sought the company of westerners during his Polynesian residence. Being in company invariably meant disagreements for Gauguin, the more so as he reached what he called his "premature old age" (Eisenman 168). Although evidence is scant, Gauguin purports to have been employed by the Department of Public Works in the capital city Papeete, from May, 1898 to February, 1899. He left his country house to his mistress while he lived in the city for work. Everything seemed to go wrong there, because time and again he felt he did not get enough respect for being a great artist. He considered his clerical work boring and humiliating, he thought an artist should be exempt from paying hospital bills, and, as if there weren't enough tension already, he entered the fray of local politics. His vituperative manner of expression made him numerous enemies. Most of the people snubbing him were Protestants, so he joined ranks with the Catholics, and expressed his vehement views against Protestants in *Les Guepes (The Wasps)* and *Le Sourire: Journal sérieux (The Smile: A Serious Newspaper)*, the latter being under his own editorship. His pen name was Tit-Oil, the nastiest sexual curse in the Tahitian language. The name of his newspaper suggests his flamboyance and the enjoyment he took in publicly denigrating others. His articles contained *ad hominem* attacks on the Governor and other Protestant officials. He used up the patience of this town before he returned home to his village, expecting to find the usual respect and adulation. Once home, he sued his wife for living in their house while he was gone. Even his old friend the Attorney General would not prosecute such a claim.

Gauguin's last isolating move was to the Marquesan island of Hiva-Oa in 1901. Eight hundred miles northwest of Tahiti, the Marquesans were thought to be the most primitive islands of Polynesia. He hoped to find peace in the "purer, more sympathetic" culture. According to Nancy Mathews, he had alienated so many people on the island of Tahiti that he *had* to move. He, on the other hand, cast himself as the agent of his own decision: "I believe that there, in savage surroundings, complete solitude will revive in me, before I die, a last spark of enthusiasm" (Eisenman 168). Tahiti was not primitive enough for his bizarre behavior; he sought a less civilized place in hopes that the primitives might overlook it.

Despite his professed desire for "complete solitude in savage surroundings," Gauguin did not quit his scandal-mongering in Hita-Oa; he actually stepped up the pace. When he found a sympathetic ear in Pastor Vernier, and an arch enemy in his neighbor, Bishop Martin, he switched his allegiance to the Protestants. As head of Church and missionary, Martin was Gauguin's chief opponent in his struggles to liberalize school policies for children. He fought for native rights in areas such as taxes, school attendance, liquor prohibition, and prostitution. He encouraged families to move four kilometers away from the Mission — the legal limit — so that their children would not have to attend school. He painted a sign on his front door, "The House of Pleasure," to taunt his Roman Catholic neighbor.

Rather than journalism, he now wrote caustic letters to government officials and private rants in his *Intimate Journals*. Not surprisingly, his campaign for native independence gained him enemies in high places. Was this his intention? Was he seeking publicity so he could die a martyr? The colonial government wanted to deport Gauguin but found it too expensive and cumbersome. He was fined five hundred francs and sentenced to jail for libel. Unlike his literary hero, Jean Valjean (*Les Misérables*), Gauguin would not have tolerated jail conditions.[6] It was lucky for him, and probably for jail wardens, that he died of complications of advanced syphilis before his jail term began.

The final isolation of this antisocial man can be illustrated by the auction of his personal assets that took place in Papeete after his death. The chief buyer of Gauguin's canvases was neither a friend nor an enemy, but Victor Segalen, a theorist who capitalized on the primitivist cachet of Gauguin's life and art. Segalen later became known for his theories of the exotic, traveling over much of the Far East to study the phenomenon and its effect upon him. Thus, Gauguin's personal effects were salvaged by one who considered him an intellectual commodity rather than a friend. Likewise, after his death, Gauguin's *vahines* were discovered to have less than friendly feelings towards him. When Bengt Danielsson researched Gauguin's life in the 1950s, Tehamana (the first *vahine*) apparently would not come forth for an interview (Mathews 213). In 1917 Somerset Maugham, author of *The Moon and Sixpence* (a fictional life of Gauguin), tried to interview Pahura, the second *vahine*, without success. If he wasn't bringing her money from Gauguin's family, she said, then why was he bothering her? (Mathews 214). Neither of these intimates of Gauguin had anything to contribute to the legend, suggesting that Gauguin's last years were ones of extreme isolation.

This brief summary of Gauguin's isolating moves is not psychoanalytical; it seeks to describe the appeal and the necessity of solitude for some artists and antisocial people. Although they seek friends and community, it is often difficult for them to sustain friendships and a good reputation. Apart from such generalities, there were several specific reasons why Gauguin sought isolation. The foremost reason is that he made so many enemies, or caused the end of so many friendships, that it became necessary for him to escape the social climate that (he believed) engendered these. In Paris, as well as Pont Aven, he made enemies of other Post Impressionists, even those whom he initially admired, such as Camille Pissarro, Edgar Degas, Emile Shuffenecker, Emile Bernard, and Charles Laval. Vincent van Gogh was an exception. His altruism, and also his unbalanced mind, account for the elasticity he exhibited towards their failed relationship. Van Gogh even wanted to live with Gauguin again after his psychotic break and Gauguin's personal break with him. On the other hand, after ten years of tolerance, support and good faith after their separation, even his wife Mette had to draw the line, citing his "ferocious egoism" that "surpasses all human comprehension" (Mathews 221). Gauguin moved to Tahiti in part to evade her claims for financial support. In his lifetime, he sent no more than a few thousand francs to support wife and children. The emotional support that he demanded and often received from her was certainly never repaid in kind.

Besides giving him the chance to start over socially, isolation in Tahiti also served to make Gauguin, if not the leader of Post Impressionism, at least the unique purveyor of the exotic dream in paint. He needed Tahiti in order to pull away from the Synthetists Emile Bernard and Louis Anquetin, whose similarity in technique resulted in debates about who started the school. Art historians generally do not consider this question germane, in part because it is unanswerable by mere comparison of the dates of canvases, and in part because, given the distance Gauguin quickly traveled from his own Synthetist style, the question becomes moot in the longer term. However, the occurrence of shared technique among the three painters serves as a clear example of the close collaboration and shared vision within several other schools of Post Impressionism (Pointillists, Nabis, Fauvists, and Neos, to name a few). The particular techniques of the Synthetist school included flattened forms, dark outlines around forms, and the "purification" of color by reducing the spectrum to "the seven fundamental colors" ("Synthetism").

Two paintings dealing with the same subject and painted contemporaneously invite comparison between Gauguin and another Synthetist. Emile Bernard's *Black Wheat* (1888–9) and Gauguin's *Haymaking* (1889) show similar, but not identical techniques. Bernard's outlines are thicker and the abstract qualities greater. Gauguin capitalized on the other two techniques of Synthetism, purification of color and the flattened form. After leaving Brittany, he continued to outline his figures as well, but in less attention-calling ways. He and other Synthetists had been accused of eliminating shading and the divided brush stroke of the Impressionists because of their inability to achieve them. This charge is easily disproved by the modeling, or shading, in Gauguin's *Portrait of a Woman* (1890), and the characteristically Impressionist divided brush stroke in *Garden in the Snow* (1883). His disavowal of shading and divided brush stroke was not from any lack of ability but from a disagreement with the ideology behind them.

However radically they broke from Classical tradition in subject matter and brush stroke, the Impressionists were still naturalists. Even though their focus was on the impressions that light made on the eye and mind, they still wanted to render what the eye saw. The Post Impressionists were anti-naturalistic. They wanted to symbolize what their hearts and bodies felt when their eyes looked at a scene or object.

In the highly competitive world of *fin de siècle* painting, isolation was important to Gauguin for finding, not only new technique, but even more importantly, new subject matter. In many scholarly fields, the primary qualification for publication of an essay or a book is whether it is relatively uncharted territory. Alternatively, it may address a familiar subject, but needs to say something new. In Gauguin's field, the same commandment applies. In his time, Impressionists had already made a seismic shift from classical subject matter. Impressionists introduced commonplace scenes and objects as viable subjects, painting what they called "a corner of the world." Classical art, on the other hand, was historical; it aimed to reproduce literary narrations (mythological, Biblical, national, military) with the greatest possible semblance of reality. For Gauguin to utilize new techniques already in use by his friends— such as flattened forms, bird's eye view, primary colors, and symbolic motifs— was one thing. But introducing entirely new subject matter — the people, customs, and landscapes of exotic Tahiti — was another matter. It was radical, and calculated to achieve notoriety. The fact that Gauguin could not find a painting companion to accompany him to Tahiti attests to the adventurousness of the move at the time.[7] No artist, however unconventional in his career choice, was willing to give up his comfortable existence in the center of the art world for the isolating, unknown experience that Gauguin proposed and executed. It was also difficult for them, as for him, to save the money required for the endeavor. Gauguin's courage and vision in this regard earned him much in the way of a reputation for individuality.

The most essential reason to choose Tahiti was to become a noble savage, which Gauguin thought was his birthright. During his Brittany years, he had begun calling himself the "savage from Peru" and claiming Indian blood, for artistic cachet. These claims were untrue; though he did have ancestors in Peru, they were not Indian. The sexual aspect to his fantasy is of course the most publicized. In a land where he dreamed that women were naked and sex was readily available, he would be sexually satisfied and socially free of the puritanical judgments of his French peers and Danish wife. But by far the more prominent feature of his dream (*pace* Mathews) was the fantasy that European greed and competition would be absent in Tahitian society. When he spoke of his choice, he stressed the non-competitive nature of society in his Polynesian idyll, not the sexual availability of women.

In his attraction to Tahiti, Gauguin very quickly developed from an *exoticist* (one who values difference for its own sake, without seeking to understand its historical and cultural causes) to a *primitivist* (one who values primitives because they live within a more egalitarian social structure than people in industrialized nations). Whereas exoticists may believe they can attain primitivism by sleeping with natives, primitivists may shun this touristic, exploitative approach. Primitivism is more of a "rhetorical stance," which uses the ideal of primitive virtue as the basis of describing a model for a happier future society (Eisenman 80). Gauguin was no political theorist, like Jean-Jacques Rousseau, who popularized the phrase, *the noble savage*, but his writings (particularly *Noa Noa*) show how far he progressed from the exoticist agenda of the bestseller *The Marriage of Loti* (Pierre Loti 1880), which was the sole contact most Frenchmen had with Tahiti. On the other hand, reverse racism is inherent in any system that elevates one race over others, as most types of primitivism do. The problem with putting anybody or anything on a pedestal is that they have far to fall when one's illusions are shattered. I will return to issues of race in a subsequent section.

Although his own stance was more complex than mere sexual tourism, Gauguin did capitalize on the exotic appeal of Tahiti to sell work. Tahiti's isolation served Gauguin well, even if his dream did not materialize quite as he had imagined. Distance provided a smoke screen from the prying eyes of his colleagues and buyers. This screen enabled him to fabricate a misleading and inaccurate Polynesian mythology with which to exoticize his canvases and books—simply because Parisians did not recognize his scholastic errors. He shrouded himself in the cloak of exoticism; to be exotic, one must be inaccessible. Gauguin cleverly used exoticism to attract the art market, even after his own interests in Tahiti had deepened to a form of primitivism.

Tahiti's distance from Europe also played a key factor in Gauguin's choice of the islands. As exotic French colonial holdings, both Madagascar and Tonkin (now Vietnam) were Gauguin's other options. Both were represented in the 1889 Paris World Fair by replicas of native villages and half-clothed "savages" who were also available for assignations. This experience definitely titillated Gauguin, and he took advantage of it. But he ultimately rejected Madagascar for being too near and too accessible to other Europeans, whereas Tonkin rejected his visa application. Being both smaller and more remote, Tahiti offered the best choice. It is the remoteness and smallness of Tahiti, the "jewel of the Pacific," that exerted, and continues to exert, its greatest appeal. In the imagination, it remains untouched by the ravages of civilization. In reality, colonization had been underway for nearly a century when Gauguin went there, and continues today in actions such as France's nuclear testing on nearby atolls in 1995–6.

"Be Mysterious": Intentional Obfuscation

One of Gauguin's more enigmatic woodcuts is *Be Mysterious* (1890). It depicts a red-haired, savage woman seen from behind. She looks up in wonder at her reflection in the sun, while a serene, dark-haired figure blesses her from behind. The words "*Soyez mysterieuses*" appear in the upper left-hand corner. Both title and imagery suggest not only Gauguin's destination — the land of sun and savages— but also his evolving artistic manifesto. Art historian Christoph Becker believes that Gauguin "gradually eliminated communication from his art" and "intended to negate all of the preconceived criteria for interpretation in order to achieve an autonomy of content" (81). Tahiti aided the process. This will become clear in my discussion of the late masterpiece *Where Do We Come From?*

In the context of using Otherness to mystify or to clarify, Darwin and Gauguin's uses of islands are nearly opposite. Darwin used the Galapagos to *clarify*, while Gauguin used Tahiti to *mystify* and *obfuscate*. Darwin used the Galapagos experience to prove facts of nature, while Gauguin used Tahiti to express personal feelings. Darwin's project was conceived as a contribution to the progress of civilization, whereas Gauguin's was an intentional regression to primitivism. Gauguin so obfuscated the realities of Tahiti that Mathews finds nothing in his canvases that resembles the island's reality. His purpose for this obfuscation was to prove his unique visionary genius. In contrast, Darwin's best work was done in collaboration with other scientists. Although it troubled him when Alfred Russel Wallace sent him a short essay delineating his own theory of natural selection, Darwin sent Wallace's work to the scientific community to which he (and not Wallace, by reason of class and background) had access, resulting in a joint publication of the seminal work. This example serves to illustrate the collaborative nature of scientific work. One person cannot possibly create an individual science. Science is achieved by centuries of data collection and theory testing, in which one must always rely on fundamentals gleaned by one's predecessors. Gauguin was a modern artist, to whom *individuality* was vital. His goal was to break away from the Synthetists and Neo-Impressionists in order to establish a new way of seeing. Like a scientist, he built upon the discoveries of his forebears, but unlike a scientist, it was crucial to his vision to be different from them, to say something new. To this end, Gauguin needed isolation. By the time he moved to Tahiti, he was no longer in collaboration with any of them, and by the last years, he was quite out of touch with what the Paris circle were doing (Mathews 211).

Degenerationism, Polygenesis, and the "Colonial Two-Step": The Uses of Social Darwinism in Gauguin's Time

Gauguin and Darwin share a more interesting and subtle tie than using isolation for different ends. Although Gauguin did not believe Tahitians were examples of *degeneration* (the gradual extinction of a species due to their unfitness), certain art critics in Paris read his canvases in this way. Francois Thiebault-Sisson's 1892 critique of the canvas, *Te nave nave fenua* (*The Delightful Land*), reveals such degenerationist thinking. The painting depicts a large, Tahitian Eve standing in a paradisiacal garden. She supply sways her hips and arms, apparently unperturbed by the hideous, winged serpent about to bite her hair. She has large, rectangular feet, just like most of Gauguin's painted Tahitians, and like real Tahitians (Henry LeMasson and Charles Spitz's Tahiti photos from the 1890s show that Tahitians did have large, healthy feet, the toes not mangled or curled from wearing shoes). By some mistake of rendition, Gauguin's Eve appears to have seven toes on her left foot. Thiebault-Sisson writes, "why has the artist so far forgotten himself as to see in the Tahitian woman of today, as in the Tahitian woman of old, only a female quadrumane?" (Eisenman 66). *Quadrumane* means a being with four hands— in other words, a simian creature that has the ability to grip objects with its feet as well as its hands. In linking this Tahitian Eve to a monkey, Theibault-Sisson suggests that she is lower on the evolutionary scale than a white woman.

Although the seven toes are clearly a mistake, even sympathetic critic Stephen Eisenman (author of *Gauguin's Skirt*) finds in this canvas an illustration of Gauguin's ambivalence about "the explanatory significance of race":

> Gauguin's *Te nave nave fenua* is thus uncomfortably located on the margin of two racial regimes. The figure of Teha'amana/Eve is neither desirable nor repellant, neither Tahitian nor European, neither classical nor modern, neither black nor white, neither Venus nor chimpanzee and neither originative nor degenerate; she is in every way a liminal character who represents but does not resolve the contradictory perspective of the metropolitan male artist toward the colonial female subject whom he wishes to know in every sense of the word. Teha'amana's curious disfigurement is the most salient sign of her liminality… [66] [Teha'amana is the name of the model and Gauguin's *vahine*].

Just as Gauguin justified the messy drawing in his finished paintings as eruptions from the unconscious that deliver a message of their own, Eisenman believes that the mistake of the seven toes is a symbol, not of Gauguin's racism, but of his "liminal" or marginal positioning with regard to colonial privilege. He elsewhere maintains that Gauguin is implicated in the powers, privileges, and perceptions of the colonist, but does not agree with them, at least not consciously. Gauguin straddles the line between colonist and subaltern, partaking of the views of both. Sitting on the fence is a difficult, but rather common stance for Europeans attracted to the exotic. Not only is it difficult for them to acknowledge their position of privilege; it is also hard to give it up. How does one begin to give up one's own privilege? In an ideal, post-colonial society, would white males truly be enjoined from representing the Other? Such a resolution, in my view, would represent a loss of art and cultural bridging.

Another explanation for Gauguin's liminality with regard to race and gender is the intellectual and social context of his times. The nineteenth century witnessed a proliferation of justifications for both slavery and colonialism by using pseudo-scientific reasoning. Apologists for racism *tried* to be empirical, calling phrenology and physiognomy sciences, but they were in fact purely theoretical in their approach to race. In the 1820s an English biologist named William Edwards formulated a doctrine of race that emphasized both the unalterability of races and their moral hierarchy. Degenerationism, a racist theory expounded by DeGobineau, B.A. Morel, Gustave Le Bon, and Max Nordau, held that racial feebleness (rather than disease, massacre, or de-culturation) caused the demise of certain indigenous communities in contact with European colonizers. When Gauguin arrived, Tahiti's population had fallen from 100,000 at the beginning of the colonial enterprise to 9,000 people. Imported diseases, alcohol, and slave trade, as well as "a psychological response to the sense of feeling hopelessly trapped" caused the reduction in native population (Heerman 164). In addition, French Protestant missionaries challenged the authority of the British in the 1840s, fomenting a civil war that decimated the native population. (Mathews 218) With such staggering figures, the main purpose of the "scientific" theory of degeneration emerges ... to assuage colonial guilt.

The title of Léopold de Saussure's 1899 book *Psychologie de la colonisation française* makes its psychological agenda clear: to condone colonial conquest by asserting the moral and intellectual superiority of Indo-Europeans. *Polygenesis*, a pseudo-scientific theory closely related to degeneration, derives its clout from Darwin's work on plant and animal species. It holds that physical racial differences, such as skin color, stature, and hair type, create as distinct a difference between humans as species do among animals. Just as different animal species cannot successfully mate with each other, neither can different races of humans. Polygenesists argued that "mixed breed" humans either died, reverted to the dominant type (copying Darwin's terminology), or were infertile. If human races were too different to mix, says this theory, then they must have come into being at different places and times and engaged (just like Darwin's species) in a competition for survival. Darwin's famous phrase "survival of the fittest" thus played neatly into a racist historical narrative, in which non–Caucasians lost in the competition for survival because of their inferiority. The dark races were considered incapable of progress due to their "arrested development, unhealthy inbreeding, or infantilism"; as less fit species, Blacks, Indians, and Asians were bound to die out (Eisenman 52).

In spite of the radically racist climate of the *fin de siècle*, Gauguin did not believe in either degeneration or the moral inferiority of darker races. Even for one as prone to contradictory impulses as himself, such a belief would be untenable, since his entire life change in 1891 was based on the belief that Polynesians were more natural and pure than Europeans. Their lifestyle was something he sought to emulate. On the other hand, he did express racism towards other groups, as when he called the Chinese "a yellow stain" in the Pacific, and depicted his friend Jacob Meyer de Haan as a Jewish caricature, with red hair, skullcap, slanted eyes, and clawed feet (Eisenman 203). Eisenman describes Gauguin's ambivalent and inconsistent attitudes toward race and gender as a colonial two-step, in which "every aesthetic and political breakthrough was followed by a setback" (204).

The colonial two-step metaphor could easily apply to Darwin's own dance on the *Beagle* voyage, as he alternated between shame and pride at colonial abuses and achievements. Darwin himself espoused a "developmental theory" of race, or at least alluded to such a belief in *The Voyage of the Beagle* when he claimed that Fuegians were racial forebears to European whites, and that South Sea Islanders were behind in social, if not intellectual, development. However, both Darwin and Gauguin's attitudes must be contextualized within their class, place, and time. The enlightenment of the privileged classes does not occur in a day or even a generation. By forcing them to confront their own prejudices and to witness the ill effects of colonialism, their journeys to islands broadened both men's views and increased their appreciation of difference in a far more profound way than mere exoticism or tourism.

Savagery and Aesthetics: What Tahiti Did for Gauguin's Painting

What Tahiti did for Gauguin's painting was in part what the various schools or movements (Nabis, Synthetists, Symbolists) did for other Post Impressionists' painting in France. It taught him ways to enact the modern art credo which held that the artist's expression of his own emotion was the primary purpose of art, to which precision and verisimilitude were subordinate. However, isolation provided Gauguin the opportunity to develop both individual techniques and new subject matter that his French colleagues lacked. I will first describe some of the changes wrought in Gauguin's painting style from contact with the Polynesian environment, then illustrate with two canvases, a "before" and "after."

Tahiti's greatest contribution to Gauguin's art is an increased brilliancy of color. Already, the Post Impressionists had exploited the dramatic effects made possible by juxtaposing the three "color complementaries," which are green and red, yellow and violet, orange and blue. Several well-known paintings illustrate the popularity of these stimulating contrasts: Van Gogh's violet irises against a yellow background in *Vase with Irises* (1890), and Gauguin's red cliff next to green pasture in *Above the Chasm* (1888), and Maurice Denis' abstract in red and green called *Sun Spots on the Terrace* (1890). Yet Gauguin goes further with complementarity; he juxtaposes both complementary and adjacent hues. Unlike van Gogh's famous complementaries of red/green and yellow/violet, Gauguin's signature hues are orange, mauve, pink, and lavender. More than the Nabis, and prior to the Fauvists, Gauguin dazzles the eye with a full range of color to demonstrate his credo that color is the expression of emotion. Color expresses how the scene makes the painter *feel*, not what color film (if it had existed) would have reproduced. In addition, Gauguin boldly introduces purely arbitrary colors (like the Fauvists), such as a vermilion horse (*The White Horse*, 1898), pink grass (*The Call* 1902), or a brilliant yellow field (*Sacred Mountain*, also known as *There Lies the Temple*, 1892).

Second, Gauguin flattened his forms even more in Tahiti than he had done in Brittany, again according to the belief that form and color, rather than precision, create tension in a canvas. The Post-Impressionists believed that modeling (shading) was cheating because it tended to hide the difficult areas of the canvas. In combination with the enhanced color and exotic subject matter in the Tahiti pieces, the flattening mechanism rendered a more exciting result there than it had in Brittany.

Third, in Tahiti, especially in the 1892 phase of two-figure close-ups, he brought his figures to the fore. In these paintings, the traditional aesthetic of figures situated naturalistically in landscape disappears. Instead, the figures fill the frame or are cropped by it, and thereby command our attention.

Fourth, in Tahiti he enhanced the decorativeness of his paintings. To a naturalist, "decorative" is a dirty word, but to a Post Impressionist, it has a positive value. Maurice Denis of the Nabi school expressed his consciousness of the materiality of the canvas in an oft-quoted statement — "a painting is a flat surface covered with colors arranged in a certain order" (Zelder 216). These colors and their order should be pleasing to the eye. Denis' attention to the two-dimensionality of the medium emphasizes the formal and decorative nature of Post-Impressionist painting in place of the verisimilitude sought by both Classicists and Impressionists. The Nabis took this appreciation of decoration further, and branched out to the applied arts—ceramics painting, lithograph making, and poster design. Gauguin tried decorative media as well, in the form of wood statues, cups, and block prints. One of the greatest appeals of Gauguin's paintings is his full, decorative coverage of the canvas, with flowers, fruits, trees, animals and streams. In interiors, he covers the canvas as well; tapestries, carvings, fabrics, toys, *objets d'art*, and light bursts decorate the fringes. As well as giving a sense of fullness, decoration provides a pleasing sense of balance in the canvas. Curved lines suggest the harmony of the natural setting and the tranquility of his mood, as least in the earlier sojourn.

Fifth, Tahiti increased the mystery and symbolism of his work, sometimes through indecipherable facial expressions (that have been compared to Mona Lisa's smile), other times by the inclusion of mystery figures, such as the *tupapau*, or spirit of the dead, that haunts the Tahitian night. He added mysterious titles with no apparent con-

***Sacred Mountain*, painted in 1892, has a second name given in the Tahitian language (Parahi Te Marae, meaning "There lies the temple"). In this canvas, Gauguin mixes icons from diverse cultures. Paul Gauguin. Philadelphia Museum of Art: Gift of Mr. and Mrs. Rodolphe Meyer de Schauensee, 1980.**

nection to the subject matter of their respective paintings. Titles such as *Parau-Parau* (*Whispered Words*), *Aha oe feii* (*Why Are You Jealous?*), *Nafea faa ipoipo* (*When Will You Marry?*) and especially *Where Do We Come From? What Are We? Where Are We Going?* puzzle the viewer. Gauguin also incorporated mysterious Polynesian religious motifs into his work. Culling them from several different cultures (Hawaii, the Marquesas, and Easter Island were favorites), and from an outdated book, *Voyages aux iles du Grand Ocean* by Jacques-Antoine Moerenhout (1837), he passed them off as both Tahitian and contemporary, not caring whether his Paris audience knew the difference. For seventy-five years prior to Gauguin's visit, missionaries had outlawed Tahitian religious and artistic traditions. When he arrived, Maohi (native) art, dress, dancing, singing and worship were virtually wiped out. Even his native informants could not be accurate about the details of a past that colonists had attempted to erase.

A good example of both the anachronism and cultural hybridity of Gauguin's religious borrowing is the painting called *Parahi Te Maras* or *There Lies the Temple* (1892). The foreground contains decorative pink and red bushes before an ornamental fence decorated with the shapes and skulls typical of Chinese art. The yellow field is pure van Gogh in his Arles period. In the middle ground and slightly to the right of center sits a large Polynesian idol-statue with sacrificial (or ritual) smoke rising in front of it. The ethnological realities of place are completely different from what is represented. The Maohi did not carve idols of their gods, their stone temples were annihilated by 1891, and they did not build Chinese fences. In Moerenhout's book, Gauguin had read about a ten-meter stone statue of the moon goddess Hina, but that was on Easter Island, not Tahiti. Gauguin was proud of his auto-didacticism, and did not bother about academic correctness. He even published his own book of Tahitian religion, *Ancien Culte Mahorie* (1893), which was largely cribbed from Moerenhout. While it does not appear that Gauguin personally adopted the beliefs of the Maohis (Mathews even calls him a distinctly unspiritual person), some of their motifs and beliefs decorate and inform several of his canvases. During the second sojourn, more canvases depict Christian themes, such as *Nativity*, *Hail Mary*, and several Eves. This does not indicate nostalgia for the religion of his nation, but rather a calculated appeal to a French art market that still adored paintings of Biblical narratives, yet was open-minded enough to enjoy their transplantation to a Tahitian setting. The hybridity of colonizer and native cultures represented by these depictions interested Gauguin as well, and he wondered in *Noa Noa* how his mistress Tehamana perceived Christianity and what role it played in her life.

Sixth, in the Tahiti works, humans, rather than animals, objects or nature, are the main forms. The female Tahitian figure fascinated Gauguin by its difference from the European average. He writes sensitively about the relationships between sexism, patriarchy, and fashion in *Noa Noa*:

> Among peoples that go naked, as among animals, the difference between the sexes is less accentuated than in our climates. Thanks to our cinctures and corsets we have succeeded in making an artificial being out of woman. She is an anomaly, and Nature herself, obedient to the laws of heredity, aids us in complicating and enervating her. We carefully keep her in a state of nervous weakness and muscular inferiority, and in guarding her from fatigues, we take away from her possibilities of development. Thus she is modeled on a bizarre idea of slenderness to which, strangely enough, our women have nothing in common with us, and this, perhaps, may not be without grave moral and social disadvantages [Eisenman 116].

The passage shows extraordinary agreement with feminist beliefs about the confining and debasing properties of much of female fashion. Gauguin appreciates the comfort and equality that the larger-limbed Tahitian woman enjoys by virtue of her freedom from European fashion. Apart from the exoticism of the final sentence, the following passage again displays a remarkably utopian and androgynous world:

> On Tahiti the breezes from forest and sea strengthen the lungs, they broaden the shoulders and hips. Neither men nor women are sheltered from the rays of the sun nor the pebbles of the sea-shore. Together they engage in the same tasks with the same activity or the same indolence. There is something virile in the women and feminine in the men.... Man and woman are comrades, friends rather than lovers, dwelling together almost without cease, in pain as in pleasure, and even the very idea of vice is unknown to them [Eisenman 116].

In addition to their natural physique, the brilliancy of the women's clothes (both the *pareus* or large scarves, and the Mother Hubbards or mu-mus) must have appealed to Gauguin's highly refined sense of color, as evidenced by the deep, rich purple dress in *Woman with a Mango* (1892), the hot pink which belies the title of *Reverie* (1891), and the pastel pink of the Mother Hubbard contrasted with the primary red of the *pareu* in *Tahitian Women* (1891). There is a vibrant color contrast in the *pareu* fabric, between the white or yellow designs and their brilliant backgrounds, which is further enhanced by its juxtaposition with brown skin.

Before and After Tahiti: Breton Catholics and a Tahitian Pagan

The painting *Mana'o tupapa'u* (*Spirit of the Dead Watching,* 1892) is considered the "manifesto" of the first Tahitian period because Gauguin re-uses it in a half dozen other pieces (Eisenman 135). It certainly expresses much of what Gauguin learned about both style and content during the first stay on Tahiti. I will compare it to the manifesto of his Brittany period, (*The Vision After the Sermon: Jacob Wrestling with the Angel,* 1888), in order to demonstrate the Tahitian developments described in the previous section.

The Vision After the Sermon is a 28 × 36 inch canvas depicting a group of white-bonneted, Breton women gathered in a semi-circle around a vision of Jacob wrestling the Angel, which is inspired by the church sermon they have just attended. Their priest appears in profile at bottom right. All of the figures are viewed from above and behind. Excepting the central woman, who is looking directly at the vision, all of the peasants are bowing with hands folded or in a gesture of prayer. A bare, brown tree trunk diagonally bisects the middle ground. A cow nibbles leaves. The vision in the right background shows two men wrestling, one with wings. By actualizing their vision, Gauguin intended to convey the deep religiosity of Breton peasant women. He borrows several techniques from Japanese woodprints, including the bird's eye perspective, the divided canvas, the relatively flat forms, the bright colors, and the form of the wrestlers (copied directly from a Japanese print).

The Synthetist influence is also strong, as shown by the black outlines around the figures, the simplified style of drawing, the rhythmic function of the line, and the highly subjective color (brilliantly red earth, probably grass). Although we know that Gauguin admired the simplicity of peasant life (he once said that the sound of his clogs on the cobblestones made him feel like he belonged), it is possible to detect his disparagement towards the simple, religious women in

***The Vision After the Sermon: Jacob Wrestling with the Angel*, 1888. There are strong Japanese and Synthetist influences in this canvas. Gauguin portrays a quite literal faith in Scripture as these Breton peasant women visualize the sermon they have just heard at church. Paul Gauguin, The National Gallery of Scotland.**

their hard expressions. The lips are frowning or pursed, suggesting a tightness or effort in the vision, which is quite different from the faces of wonder and mystery on their Tahitian counterparts. The mere fact that the nine women and their priest are having the same vision may also suggest their lack of imagination or the social conformity of their type.

Mana'o tupapa'u (*Spirit of the Dead Watching*), 1892, is also a 28 × 36 inch painting. It depicts a young Tahitian girl (Gauguin's *vahine*, Teha'amana) lying prone on a bed covered by a yellow sheet draped over a bright blue and orange *pareu*. Just behind the bed sits a rigid, darker-brown figure in profile, dressed in black cowl and cloak, against a decorated pole. The purple background, which is either the

Spirit of the Dead Watching, 1892. **Possibly his most famous canvas of the first Tahiti period, this painting derives from a moment in which Gauguin frightened his *vahine* by coming home late from town. He imagines the vision she might have had, of the night spirit lurking beside her bed. The painting has spiked controversy throughout the century because some interpret it as imperialist and sexist. Paul Gauguin, oil on burlap mounted on canvas. Framed 36¼ × 44½ × 2½" (92.075 × 113.03 × 6.35 cm), Albright-Knox Art Gallery, Buffalo, New York. A. Conger Goodyear Collection, 1965.**

back wall or the night outside, is decorated by bursts of white light, blue leaves, and a brown arc (perhaps part of a decorated headboard, but this is indecipherable). A bright pink pillow or dress under the girl's yellow pillow completes the symphony of color. In his book *Noa Noa*, Gauguin tells the story behind the painting. Coming back late at night, he surprised and frightened Teha'amana in her bed. In a letter to his wife, Mette, Gauguin explained that he had wanted to paint a "slightly indecent" Tahitian nude, but that he felt he needed to add a story to make it palatable to Paris. That story was Tehamana's spirit vision; she mistook her lover Gauguin for a spirit of the dead.

It is curious that art historians have reacted so strongly to this picture. Although the audience of his time might have been unused to full dorsal nudity, there is nothing indecent about this painting today.[8] The girl is attractive, but not sexy — she has the androgynous square hips that Gauguin admired. The look of fear that is supposed to provide the erotic touch is barely perceptible (she looks merely sleepy). Her legs are crossed, her breasts and genitalia hidden. Yet Nancy Mathews reads a whole history of sexual abuse into it: "Gauguin's previous battering of Mette suggests that despite his innocent rejection of Tehamana's pleas to punish her in *Noa Noa*, he very likely did abuse his Tahitian vahine, the submissive fear in her eyes being his erotic reward" (182). Even Eisenman, usually more even-handed than Mathews, calls the picture and the written account of it "a veritable encyclopaedia of colonialist racism and misogyny" (120).

Both scholars use Gauguin's own words, rather than the painting, to condemn him. Each scholar reads selectively, omitting important sentences that describe Gauguin's concern, his perception of Tehamana's Otherness, and his love of her for that very reason. Gauguin's passage reads:

> I saw [Teha'amana] ... immobile, naked, lying face downward flat on the bed with the eyes inordinately large with fear. She looked at me, and seemed not to recognize me. As for myself I stood for some moments strangely uncertain. A contagion emanated from the terror of Tehura. I had the illusion that a phosphorescent light was streaming from her staring eyes. Never had I seen her so beautiful, so tremulously beautiful. And then in this half-light which was surely peopled for her with dangerous apparitions and terrifying suggestions, I was afraid to make any movement which might increase the child's paroxysm of fright. How could I know what at that moment I might seem to her? Might she not with my frightened face take me for one of the demons and specters, one of the Tupapaus, with which the legends of her race people sleepless nights? Did I really know who in truth she was herself? The intensity of fright which had dominated her as the result of the physical and moral power of her superstitions had transformed her into a strange being, entirely different from anything I had known heretofore [*Noa Noa* 33].

In characterizing Gauguin as abusive in this moment, both Eisenman and Mathews ignore several factors, particularly his own fright and uncertainty. His appreciation of her beauty is not expressed as sexual, but spiritual (light in her eyes). Her beauty does not have to signify that she is his sexual victim. His words do not intimate that her fear is a turn-on because it gives him power. On the contrary, he freezes so as not to frighten her further. The art historians' readings are highly selective and political.

When read in conjunction with a later passage in the same book, Tehamana's look of beauty and strangeness at 1 a.m. can be understood as an imagined spiritual transformation before which Gauguin stands humbled rather than empowered. The later passage takes place after Gauguin has spent a day of fishing with the natives. They relay a folk belief which holds that if the hook enters the fish's lower jaw, the *vahine* has betrayed her *tane* (husband) that day. Both Gauguin's catches were lower-jaw hooks. Upon confronting Tehamana with the question that evening, Gauguin undergoes the closest thing to a religious experience ever described by him or attributed to him by anyone else. Her "extraordinary expression of mysticism and majesty and strange grandeur ... transformed the atmosphere." He writes:

> *I felt that something sublime had risen up between us.* In spite of myself I yielded to the influence of Faith, and I was waiting for a message from above. I did not doubt that this message would come; but the sterile vanity of our skepticism still had its influence over me, in spite of the glowing sureness of a faith like this rooted though it was in some superstition or other [his italics] [*Noa Noa* 63].

In these sentences, her faith so moves him that he seeks to believe, if not in the superstition about fish, at least in some higher power. After saying a prayer, Tehamana asks Gauguin to strike her many times. Again, he sees something sublime in her look, as when he had surprised her in the dark: "In the profound expression of this face and in the perfect beauty of this statue of living flesh, I had a vision of the divinity herself who had been conjured up by Tehura [another name for Tehamana]" (64). Far from striking her, he murmurs a quotation from Buddha, "By kindness you must conquer anger; by goodness evil; and by the truth lies." (*Noa Noa* 64). With this thought, he is instantly healed of his jealousy. While readers familiar with his life will certainly question the sincerity or purity of Gauguin's protestation of innocence, it appears that the occasion of *Manao Tupapau* was a sense of Tehamana's spiritual and physical beauty, rather than a desire to molest or abuse her.

To return to the comparison of *The Vision After the Sermon* with *Manao Tupapau*, we see that thematic content links these two paintings. In each one, women are apparently having visions ("apparently" because Gauguin stresses that he does not know what Tehamana is thinking). What is strikingly different is their composition. In the earlier painting, the protagonists face their vision (except for three who are inexplicably facing the viewer of the painting), whereas Tehamana faces the viewer and looks away from her vision. A more telling difference is that the Breton women's vision had been supplied though the words of the priest, whereas Tehamana's vision is mysterious to Gauguin, and unmediated by any interpreter. He can only guess what the *tupapau*, or spirit, looks like. In *Vision*, the congregation was told what to think of the event; in *Manao*, Gauguin had to imagine, not only the appearance of the spirit, but also the feelings it might elicit in Tehamana. These differences point to the greater degree of mystery and exotic symbolism in the Tahitian period. Gauguin appears to be yielding authority to the Other to suggest its own story.

Remember that in the Tahitian period, Gauguin began to foreground the focal point. Whereas in *Vision*, the center of interest is the wrestling match in the background, in *Manao Tupapau*, it is Tehamana, who appears in the foreground. This foregrounding of the focal point energizes and dramatizes the canvas. Whereas Gauguin places his simple, conformist Brêton women with their backs to the viewer, he places his beautiful Tehamana with her face to us. Her expression, directed at the viewer, is the focal point of the canvas. The whole format suggests the greater simplification and daring of the Tahitian years.

The two canvases nicely demonstrate the enhanced decorativeness and brilliance of color in the Tahiti works. Of course, the bright red ground of *Vision* is a radical departure from the soft hues of Impressionism. Gauguin also employs the complementary contrast of red-green with the short strip of green grass at the very top of the canvas and the green of the nearest woman's shawl. But *Manao* employs the less frequent, and more harmonious complementaries of orange-and-blue in the *pareu*, and yellow-and-violet in the sheet and backdrop. The dark brown body against the acid yellow sheet is richer than the black-and-white contrasts of the women's bonnets and bodices in *Vision*. The decorative touches of the Tahiti painting are obvious. The *pareu*'s orange flowers, the pillow's red-and-white buds, the mysterious, free-floating, blue leaves and white-light bursts, and the decorated pole in pink, blue and red all constitute decoration, an attribute that is absent from the Brittany canvas.

Turning to form, we see that the Tahiti canvas is more two-dimensional. Whereas the line of nine women in *Vision* recedes to the distance with the characteristic foreshortening of the naturalistic perspective, the spirit figure behind Tehamana in *Manao* is smaller, not for reasons of perspective so much as for theme. As Gauguin explained in his letter, he had just wanted to do a Tahitian nude. The spirit figure, *Tupapao*, is the afterthought, whose rendition is less important to him than the model's. The spirit's profile is highly stylized and flat, compared to the rounded limbs of Tehamana. It is hard to tell just where the spirit is located; such spacelessness, or flatness, is typical of his Tahitian works.

One final example of the disavowal of naturalistic perspective is the hands, which are disproportionately large even if we grant that Tahitians' hands and feet *are* large. The feet and the whole bottom edge of the bed are inaccurate, as the feet seem to be sliding off the bed toward the viewer, while the bed just drifts off the edge of the canvas with no clear edges. Perhaps, as Gauguin was to say about his later masterpiece, *Where Do We Come From?*, the mathematical faults are necessary, in the sense that precision kills the emotional impulse which makes a good painting. Compared to *Manao*, *Vision* is a mathematical picture.

Tahiti I and II: From Beauty to Allegory

Although the contrasts are not as pointed in all of the paintings, the differences between some of the paintings of the first and second periods (hereinafter Tahiti I and Tahiti II) are so remarkable as to suggest another emotional change in the painter. Before comparing the manifesto paintings of the two periods, I will suggest reasons for the emotional downturn so evident in the darker coloring, more solemn subject matter and even the depressing titles of Tahiti II (*Sacrifice*, *Self-Portrait Near Golgotha*, and *Nevermore*). Christoph Becker attributes the melancholy tones of the later period to Gauguin's sense of failed mission during and after the Paris hiatus in 1893–5. He could not work well in Paris, he did not greatly improve his financial condition, his paintings were not selling, and he gave up on communicating his message of noble savagery, considering the Parisians too close-minded to receive it (Becker 80–1). Mathews attributes the downturn to his own personality, which was hardening into more rigid lines of arrogance, deceit, manipulation, and violence that had previously been leavened by optimism and the occasional generous impulse (211). She also speculates that Gauguin was "not sensual enough to support a long-term erotic experience" (211).

Gauguin is not alone in losing the *frisson* of first encounters after long exposure to a place. It is next to impossible for anyone to make a second journey to an exotic place and find similar stimulation. Difference tends to collapse, even though mysteries remain unsolved. Part of the depression of one's hopes and dreams may be caused by the unattainability of answers, bringing with it the realization of one's own inability to assimilate. Assimilation is opposite to titillation or stimulation, as the earlier discussion of exoticism suggests. Nor am I suggesting that Gauguin deceived himself that he could "become" Tahitian, but he did hope to find a continued welcome there.

Tahiti I had idyllic qualities that are the more pronounced when compared to Tahiti II. The first *vahine*, Tehamana, was apparently more congenial than the second, Pahura, or the third, Vaeoho (Marie-Rose) in Hiva Oa. Mataiea, his village of residence during Tahiti I, was more remote and "pure" (presenting fewer opportunities to get in trouble) than Papeete, Punaauia, or Atuona, the three towns he lived in during Tahiti II. Not surprisingly, Gauguin fared better in the country than in either village or city, because he focused more on painting and less on politics. As some of his closest friends observed, his was a personality that thrived on opposition. He needed to make enemies in order to construct his persona of misunderstood genius. This self-pity did not make him any happier but was compulsive, especially in the last five years. In addition to self-destructive psychological urges, Gauguin's physical health was deteriorating rapidly. He did not admit it was syphilis that made him spit blood and caused painful sores on his legs (which could have passed for leprosy or eczema to observers) until the final move, to Hiva Oa. Characteristically cynical, he joked about his venereal disease. In *Avant e après* he wrote, "The syphilitic and the alcoholic will perhaps be the men of the future" (Mathews 254). Presumably he spread his disease to the young girls who visited his "*Maison du Jouir*" (House of Pleasure) in Atuona, but we have no solid evidence either way. Although companions of the Brittany days regarded Gauguin as one who could hold his liquor, he increased his intake in the later years to combat both depression and the pain of his legs.[9] He also took arsenic, laudanum, and morphine. He suffered a series of heart attacks between 1897 and 1903, which are now recognized as symptomatic of the tertiary phase of cardiovascular syphilis (Mathews 223). So indeed there were reasons for the canvases to darken in the last eight years of his life.

Where Do We Come From? What Are We? Where Are We Going? (1897)

According to Gauguin, this is the masterpiece, not only of Tahiti II, but of his entire *oeuvre,* though some art critics disagree. His largest canvas, at 55 × 148 inches, it is also his most ambitious subject matter, as indicated by the existential questions printed in French in its upper left hand corner. The impetus for this painting was a failed suicide attempt in December of 1897. He intended it as his Last Will and Testament to the non-comprehending art world. The fact that it cannot be understood without reference to his explanation has been a frequent criticism of it and some of the other late works.[10]

Its genre — allegorical narrative — echoes one of Gauguin's

favorite artists, Puvis de Chavannes, as well as classical French and Italian painters. Strangely enough, the canvas reads from right to left, starting with a dog and a diapered baby. Curiously, one of its three attendants is watching the baby. Two of them gaze somewhat provocatively at the viewer, while the third—naked and considerably darker—has its back turned but its face in profile. This is the first fat Tahitian Gauguin has drawn, and it has a moustache and long hair. By its body shape, however, it could be either male or female. To the left of and behind the pretty women, two gowned females walk in the shadows, whispering and gesturing as though they are engrossed in a serious conversation. Another fat, naked person (sex unknown) sits with its back to the viewer, looking at the fully-clothed pair while partially covering its face. To the left is the central figure of the canvas, an androgynous figure dressed only in a loincloth and reaching overhead to pluck an apple. It does not appear to have breasts, but the possibility exists that it is a woman whose breasts are flattened by the stretch. Nor does the garment reveal its gender, since two of the seated females wear loincloths as well. Next to the androgyne, a seated girl eats an apple. A goat, two cats and a bird (of paradise?) surround this child. A pale-blue female Hawaiian idol stands in the background. A worshiper is turning away from it. In the left foreground are seated a beautiful, though skeptical-looking woman, and a dark-skinned, withered, old lady. Finally, in the left-hand corner of the canvas, a goose is stomping on a lizard. The painting starts with a dog and ends with a goose, and is intended to represent the human life span. Gauguin says the goose represents the futility of words, yet thankfully he has provided some to explain the allegory's meaning:

> Where are we going? Near the death of an old woman. A strange stupid bird concludes. What are we? Day-to-day existence. The man of instinct wonders what all this means. Where do we come from? Spring. Child. Common life. The bird concludes the poem by comparing the inferior being with the intelligent being in this whole which is the problem indicated by the title. Behind a tree are two sinister figures, shrouded in garments of somber colour, recording near the tree of science their note of anguish caused by this science itself, in comparison with the simple beings in a virgin

> nature, which might be the human idea of paradise, allowing everybody the happiness of living [Letter to Morice, in Eisenman, 143].

This explanation is stock romanticism, shared by thinkers from Rousseau to D. H. Lawrence: Christianity and western civilization have destroyed man's happiness by too much knowledge, both moral and scientific. Gauguin explains the look of the crouching figure near the sinister ones as "astonishment at these two people who dare to think of their destiny" (Goldwater 110). True to his primitivism at the end, Gauguin valorizes the innocence of "savagery" over the knowingness of western civilization.

How does this subject matter differ from that of *Manao Tupapau*, and does the difference reflect a greater maturity in the artist during the Tahiti II period? Always skeptical of allegory (since it derived from a tradition of formalism, precision, and scholarship), Gauguin denies that *Where Do We Come From?* is an allegory. Yet the painting clearly refers to ideas and traditions outside the picture itself. In fact, he compares it to the Bible: "I have finished a philosophical work ... comparable to the Gospels. I think it is good" (Goldwater 110). *Where Do We Come From?* re-writes the *Garden of Eden*, making the first human both single and androgynous. Knowledge is punished, not by expulsion from the garden, but by the shrouded figures' relative obscurity and alienation from the healthier people in the Garden. This allegory says that the innocent life of the primitive is better than the sophisticated life of civilized man. In contrast, *Manao Tupapau* does not cast any such judgment, or for that matter, judgment of any kind. It is anecdotal, as are most of the light-splashed, two- and three-figured canvases of Tahiti I. Although Gauguin prided himself on progressively simplifying his art, *Where Do We Come From?* has quite a lot to say. In contrast, *Manao Tupapau* is an uncrowded canvas rendering the beauty of color and the human form.

Since I consider brilliancy of color Tahiti's prime gift to Gauguin, it is disappointing to observe it diminishing in the later work. The background of *Where Do We Come From?* is dark-to-medium greens and blues. The mountain, some bushes, and the cave from whence the "Christian" figures emerge, are all black or gray. The

Where Do We Come From? What Are We? Where Are We Going?, 1897–1898. **Gauguin thought of this painting as a manifesto of his belief that the primitive ways of early Tahiti, represented by partially-clothed, well-lit figures, were more joyful than the western, Christian ways, symbolized by the two sinister figures stolling in darkness. The picture and its title well illustrate his motto, "Be mysterious." Paul Gauguin, French, 1848–1903. Oil on canvas, 139.1 × 374.6 cm (54¾ × 147½ in.). Museum of Fine Arts, Boston. Tompkins Collection, 36.270.**

brightest light is cast upon the androgyne and the "astonished" woman, on whose back there is a remarkable *chiaroscuro* effect (given his penchant for flat, unmodeled figures). The two most feminine women have orange skin, not quite as brightly lit as the androgyne's and astonished woman's, but apparently placing them in the same symbolic category (the healthy ones). The child is eating an apple, perhaps symbolizing knowledge. Knowledge of the wrong kind puts her in shadow, despite the two sources of light from opposite sides of the canvas. One of these sources is the sun, represented by the gold arc containing the title, and casting a sharp shadow of a bare-limbed (and quite Japanese) tree upon the pink ground. The other light source highlights the central figure and the astonished one, perhaps suggesting moonlight. The moon's association with emotion and mystery fits well with the painting's valorization of the primitive. Apart from the two brightly lit figures and the patch of pink ground, the time might be evening. There is a sense in this painting that Gauguin's motto, "Soyez mysterieuses," is quite literal, as both the figures and their meanings are cast in semi-darkness.

In contrast, the Tahiti I period features brilliantly and consistently lit scenes. Even in the nighttime setting of *Manao Tupapau* we see a consistency of light, an absence of substantial shadow. Tehamana, the *tupapau*, and the *pareu* contain the darkest tones, yet the girl and the cloth are brilliantly highlighted with orange. Even the dark spirit's bright, staring eye calls attention. The beauty of color and form are the impetus to paint *Manao*, whereas in *Where Do We Come From?*, beauty is secondary to allegorical content. The shadowy tonality of *Where Do We Come From?* recurs in other late paintings—*Nevermore* and *Repos* (exceptionally brown paintings), and *The Sacrifice*—but the final years are not exclusively dark by any means. In two late paintings, *The Call* and *Riders on the Beach* (both 1902), Gauguin experiments with brilliant pink and orange in weird places. Pink ground is a harbinger of the Fauvists' celebration of arbitrary brilliant colors.

I began this chapter with queries about Gauguin's motives and ideologies for the Tahitian experience. I will revisit these to summarize my conclusions.

(1) How did Tahiti help Gauguin achieve his own style? Working in the milieu of Parisian Post-Impressionism, Gauguin found himself in a competitive environment. The move to Tahiti helped free his creative urge by isolating him from other artists, by exposure to colors he had not previously seen in nature or dress, by exposure to Polynesian mythology, and by relatively stress-free sexual relations.

(2) What did primitivism mean to Gauguin? Primarily, primitivism meant a freedom from money struggles (which he did not actually find), but there was an ideological component as well. It meant seeking Polynesians' noble virtues, real or imagined, such as communal sharing; less division of labor; less envy, greed and jealousy. In seeking these attributes in isolated places, Gauguin joined a long tradition of Romantics, including Jean-Jacques Rousseau, Denis Diderot, and Pierre Loti.

(3) How is primitivism different from racism? Primitivism inverts the pyramid of racism, so that darker races are on top, and Whites are at the bottom. Since hierarchy is still posited, there is still a problem of inequality. Post-structuralist literary theory proclaims that until binaries can be broken by third terms, hierarchies will

always occur, since they are inherent in any human conception of a binary, and binaries are inherent in language itself. However, there are many varieties of primitivism, and Gauguin's was not excessively racist. Unlike Pierre Loti's literary hero, Gauguin did not find an unbridgeable gap between himself and the Other that was Tahiti. He admired the spiritual qualities of his *vahine*. Another positive value of primitivism is contemporary Tahitians' use of the concept to foster Maohi cultural pride, pan–Pacific solidarity, and resistance to French colonial domination (Eisenman 202).

(4) Why was the first sojourn more successful than the second? Gauguin's expectations of playing a symbolic role as Returning Hero in 1893 were thwarted by misunderstanding of his paintings and his failure to sell them. He returned to Tahiti the second time as an exile, not an explorer. Ill health, contentious behavior, scandal-mongering, alcoholism, and a deeper isolation than perhaps even he could bear made his existence occasionally hellish, precipitating one suicide attempt in 1897 and an attempt at repatriation in 1902. Ironically, his dealer Monfried told him not to return, as Paris could only bear his eccentricities at such a distance. The artist-legend that both had carefully constructed could only serve as a marketing device as long as he was very far away (Eisenman 173).

(5) Why, in a field so dependant on collaboration, so characterized by cultish groupings, did isolation from the modern art world help Gauguin create his own best style? Precisely because Post Impressionists' works were so derivative of one another and of their older masters, Gauguin needed to leave in order to thrive. He also needed to live down rumors that he copied Synthetism from Emile Bernard, to put behind him the tragic association with van Gogh, to find new subject matter that in and of itself would say something new. He *knew* that Paris wanted exotic customs, figures, landscapes, and colors. What his audience did not accept was his symbolic mysteriousness, yet this was precisely what gave him private satisfaction and posthumous success.

(6) How did his island experience compare to Darwin's in terms of its contribution to his own discoveries? His island experiment was more different from than similar to Darwin's. His intent was to isolate himself from colleagues, while Darwin, as the nature of science

dictates, had to work with other scientists and build on their foundations. Unlike Darwin, Gauguin used distance and isolation to obfuscate and mystify his agenda, because he believed that good art comes from the unconscious. However, what is similar about islands in the two men's experience is their open ground and vital promise for vision. Although only one was an artist, both innovators used islands as canvases for their creative imagination. Islands can assume extremely visual and myth-making properties in the minds of creative people, like wide-open maps for conscious and unconscious problem solving.

Perhaps the "limited resources" characteristic of islands helped both science and art. Limited resources increased the struggle for survival and the rate of speciation, making Darwin's vision far more apparent. Tahiti's limited resources helped Gauguin to focus on what he wanted his representations of natives to say. He could change beaches or settings, but would always return to those "gingerbread" men and women in their oft-repeated postures (he used the same forms in multiple paintings). If he did not always communicate clearly in his canvases, does that mean he failed, as either primitivist or painter? Not at all. He achieved grandly — so well, in fact, that when learning the basic tenets of Post Impressionism — and thereby the foundations of modern art — a student could do no better than to study his paintings. His anti-naturalism, his celebration of color and line, his adoption of non-western motifs and techniques, his valorization of the unconscious as source of art — all these served as guidelines for both his artistic contemporaries and for future artists.

IV

"Those Three Lonely Rocks": John Millington Synge and the Aran Islands

Connections with Other Island Seekers

Although worlds apart in temperament, lifestyle, and ethnicity, Gauguin and J. M. Synge's experience of islands dovetails in important particulars. As Tahiti gave Gauguin new vision from direct use of the raw materials he found there, so Aran gave Synge the raw materials of folk culture and language from which he forged a new type of drama. The distances of each man's island from his home differ hugely. Tahiti is 9,765 air miles from Paris, while Inishmaan is about ten miles from Ireland's west coast, and only 140 miles from Dublin. The Aran Islands were closer to the metropolitan center which provided Synge's audience than Tahiti to Paris, yet this proximity only served to heighten Irish theatergoers' expectations that the plays would give a realistic (and flattering) view of west Irish peasants. In contrast, Parisians expected exoticism of Tahiti and the prodigal Gauguin. Nonetheless, Gauguin's work and life spiked controversy both before and after Tahiti. Synge generated an equal buzz in his day. In the field of Irish Studies, this bias against Synge continues today in a slightly different vein, when nationalist literary critics excoriate him as an "ascendancy parasite"[1] who appropriated folk culture for his own "self-serving" studies (Kiberd, *Inventing Ireland* 174).

Map of Ireland

Places figuring in Synge's life are named. The Aran Islands are just about ten miles off the coast of Galway County. North of Galway-town is Connemara, where Synge researched social conditions and wrote a commissioned piece for England's *Manchester Guardian.* Synge visited the Great Blasket Islands in the southwest but preferred Aran's spirit of place. County Mayo, north of Galway, is the setting of *The Playboy of the Western World.* From *J. M. Synge and His World* by Robin Skelton, Thames & Hudson Ltd., London.

Synge's connections to Darwin are less direct, and have to do with influence rather than any substantial similarities in their island projects. Born in 1871, Synge was historically and temperamentally poised to appreciate Darwin profoundly. Synge attributes the birth of his agnosticism and his rejection of his mother's Church of Ireland Evangelicalism to a passage in *The Origin of Species* detailing similarities between human hands and bat wings. At fourteen years old, his initial reaction was one of horror: "incest and parricide were but a consequence of the idea that possessed me" (McCormack 41). However, his eventual acceptance of the fact that he was agnostic led to an openness towards, and possibly spiritual consolation in, the folklore and fairy tales of the west Irish peasant. This is not to say that Synge ever believed in fairies, but that he admired the simple faith and powerful imagination of those who did. While there were many aspects of Christianity that perturbed him, Hell was by far the worst. Fitting for one who would seek escape from modernity in islands, he compared missionary zeal unfavorably with primitive rituals. "The extraordinary cruelty of thrusting the idea of Hell into the imagination of a nervous child," he wrote, "has probably caused more misery than many customs that the same people [evangelicals] send missionaries to eradicate" (Roche 110). A delicious sense of irony emerges.

Darwin influenced Synge in subtler ways as well. Whether it was part of his conscious intention in choosing Aran or not, Synge later became aware of the "laboratory" nature of islands—their isolating property gave definition and explicit boundaries to their culture. In his role as ethnographer, Synge chose a place that was more accessible to study because of its socially isolated nature. Isolation molded and retained a different culture on Aran to that of mainland communities. Like Darwin in the Galapagos, Synge in Aran assiduously observed minor but important differences in the "specimens" on the three closely related Aran Islands. Of the three islands and their populations, he preferred the people of Inishmaan, as "Gaelic is more generally used, and the life is perhaps the most primitive that is left in Europe"[2] (*The Aran Islands* 10). Also like Darwin, he hypothesized certain direct effects of environment upon character and physique: "This continual danger [of landing or launching the curagh[3]], which can only be escaped by extraordinary personal dexterity, has had considerable influence on the local character, as the waves have made it impossible for clumsy, foolhardy, or timid men to live on these islands" (*The Aran Islands* 49). Like birds with beaks too round for seed-picking on Galapagos, clumsy men are eradicated by natural selection on Aran.

A Keenly Developed Sense of Irony

Synge was not a thoroughgoing primitivist. Certainly he preferred cultures that had evaded the ravages of modernity, but he did not wish to stop time nor did he think that he could join the islanders as Gauguin tried, rather unsuccessfully, to do. His interest in folklore was aesthetic, even though motivated by a spiritual attraction to the pagan concept of harmony between man and nature. The conditions of west Ireland at Synge's time afforded many primitive places to choose from, both as balm to a tortured soul, and as fodder for his literature. The Great Famine of the 1840s killed one million Irish people, and caused another one and a half million to emigrate. The *Gaeltacht*, or Gaelic-speaking regions, lost the most people to famine and emigration, so that the Irish language dwindled drastically; after 1851 only a quarter of the population spoke Irish (Kiberd, *Inventing Ireland* 21). Why, then, were not Connemara or Kerry, both of them poor, rural, coastal counties, suitable places for Synge to settle? Why did it require an island to transform him, and why, of all the rocky outposts on the Atlantic coast, should it be Aran, and not the Blaskets,[4] where he also stayed? Critic Alan Price believes that the Blaskets did not engage him emotionally, in comparison to Aran, because the risk of death at sea was not as great (97). In his collection of essays entitled *West Kerry*, Synge noted that no Blasket man had died at sea for forty years (512). The Aran Islanders' constant struggle with the sea may not have made them great in anyone else's eyes, but in Synge's imagination, the daily *agon* was a perfect setting for tragedy as well as for his own spiritual autobiography in the form of *The Aran Islands* (published in 1907).

Furthermore, he related more to a group which, although given to occasional violent outbursts, did not brawl drunkenly like the Blasket Islanders, whose every excursion to the fairs or sports on the mainland ended in fist-fighting and kicking. They had a greater dependency on whisky and porter; Synge said there was nary a conversation that didn't include reference to alcohol. Biographers also interpret an incident in his host's cottage on the Great Blasket as an indication of its inhabitants' greater vulgarity vis-à-vis the Aran Islanders. Synge slept in the same room as his host and son, so witnessed the following. When the man awakes at 6:00 a.m., he shouts for his young daughter to serve them whisky and pipes in bed. The people of Aran appear relatively gentle, whereas the coastal people are known for their wildness.

There are several features of Aran's insularity (I use the term in its double sense—first, of simply being an island and second, of being isolated from mainland culture) that make it the place of places for Synge's transformation as man and writer. First, insularity preserved both its use of Gaelic as first language of older generations, and its particular Hiberno-English dialect. Synge studied Irish with passion at Trinity College, but never fully appreciated both its own spoken beauty and the metrical quality it provided to Hiberno-English until he lived at Inishmaan. The exaggerated expression and strong emotions of the islanders are the very ingredients of literary comedy and tragedy, a fact that Synge profoundly understood.

Second, insularity preserved old folkways and beliefs. Declan Kiberd goes so far as to say that Aran culture was pre–Christian, but this ignores the role of the priest and the prayers of the people. In particular, Synge felt attracted by their notion of after life, which was not the place of torture and atonement that so alienated him from Christianity. While the fairies were not benign, neither were they absolute in their qualities or intentions. Their other world was at times accessible to the living, so that traffic back and forth over the boundaries of Death could and did occur.

Third, the resistance to modernity, or its sheer inaccessibility, gave to Aran Islanders' lives, in Synge's opinion, a richer imagination. He spoke of them as living a dream, close to nature, and very much in concordance with its laws. This perceived oneness with the universe may have helped Synge to heal his inner divisions. Anthony

Roche and Rosemarie Battaglia write convincingly of the beneficial effects that the Aran Islanders' successful blending of Christianity and paganism had upon his own (mostly internal) fight with the Church and with his strict upbringing by an Evangelical mother.

The Aran experience, if not necessarily healing, forged a new consciousness as man and artist. Aran Islanders' unselfconscious attitudes towards sex, their subordination of sex instincts to family, and their acceptance of Synge strengthened his ego. Once a timid, reserved man,[5] Synge gained enough confidence to become a co-leader of the Irish National Theatre, fiancé of one if its lead actresses, and the playwright so sure of his own style and message that he courted the backlash of Catholics and Protestants alike. During and after the Aran visits, his writing style matured from a sentimental, self-obsessed Decadence to an other-directed, energetic Modernism, in both the comic and tragic modes. I will argue that observing Aran Islanders' stoic acceptance of their difficult life on small islands created in him the ability to see life ironically. He translated this irony into literature of stunning simplicity (on its face) and ambiguous meaning (in its depths)—a hallmark of greatness. The greatest quality of modern literature is its irony, and the challenge to the writer is to create a signature style of irony. Far from being a sentimental exoticist or a mystical primitivist along the lines of his earlier models, Pierre Loti and Anatole Le Braz,[6] Synge combined his nostalgia for the imagined past with an acceptance of that loss. From this acceptance comes the irony which ultimately blends comedy and tragedy in such a way as to celebrate life in the midst of loss, and conversely, to see human smallness in a compassionate way.

The Irish Literary Revival and Synge's Leadership

The desire to resurrect an Irish literature devalued by seven centuries of English domination[7] culminated in a movement that called itself a Renaissance or Revival as early as 1890. One branch of this movement was fervently nationalistic, and therefore in natural sympathy with two other movements of the time, Home Rule and the Gaelic League (which advocated the imposition of Irish as first language of

Ireland). Douglas Hyde and George Russell (AE) are two of the earliest enthusiasts who combined politics with literature. Hyde was a philologist who translated folk and love songs from the Gaelic. David Kiely calls him "an atavist of the first order, going so far as to advocate the abolition of trousers for Irishmen, believing (probably erroneously) that they are not of Celtic provenance" (37). AE (Moore's mystical name) was the spiritualist of the movement, though Yeats would later dabble in the occult as well. Studying theosophy and Celtic mythology, AE believed that the Revival would entice the old gods back to Ireland (Kiely 38). One of the old gods certainly did return, to literature at least, and that is Cuchulain, the hero of the Ulster Cycle of medieval Celtic legend. Every single author in the Revival incorporates Cuchulain, either with straight-on admiration or with irony. Synge's approach to the Cuchulain-worship of the nationalists is ironic; for him, the name is a code word for what is most deluded in the movement. In response to charged debates about the Irish National Theater's fare, he writes to his friend Stephen MacKenna, "I do not believe in the possibility of a purely fantastic, unmodern, ideal, spring-dayish, Cuchulainoid National Theatre, because no drama — that is to hold its public —can grow out of anything but the fundamental realities of life which are neither modern nor unmodern, and, as I see them, are rarely fantastic or spring-dayish" (McCarthy 165–6). He derided the sentimental idealism of his Cuchulainoid companions.

Just as there is division *within* the literary revival, there is dissension *without*, among the consumers of its products. William Butler Yeats, Lady Augusta Gregory, and John Millington Synge establish the Irish National Theatre in 1899,[8] a society that will produce plays by Irish writers about Irish themes, with one important qualification: though national, the plays will not be nationalistic. Thus, the gauntlet is laid down for the conflict that will ensue between the society's Anglo-Irish writer-producers and its largely Catholic — and nationalistic — audience. Synge's first-produced play, *The Shadow of the Glen*, arouses indignation because it portrays a loveless marriage and a strong woman who is aware of this fact; she talks with every tramp who passes her lonely cottage. Rather than portraying Irish peasantry in a romantic, pastoral light, Synge shows the "fundamental realities of life which are ... [not] fantastic or spring-dayish." The nationalists take this as a slur on the spiritual purity of the Irish peasant and the sexual purity of Irish womanhood. A print battle in the local newspapers is one thing, but with the production of his play *The Playboy of the Western World* in 1907, the battle becomes physical. Riots, including shouted invective and physical rowdiness, take place in the Abbey Theater, and W. B. Yeats calls in the Royal Irish Constabulary (British, of course) to keep peace in the house. The objections are the same as they had been to *Shadow of the Glen*, except that this time, nationalists' reaction comes from a deeper, unconscious place. The playboy's parodic resemblance to both Cuchulain and Charles Stewart Parnell[9]—particularly insofar as Parnell was martyred and divested of his leadership by Catholics—calls up the audience's Jungian Shadow (their repressed sense of their own faults). They both identify and resist identification with the bloodthirsty crowd, which first glorifies, then denigrates, betrays, and finally tortures the playboy, who is among other things a Christ figure (McCarthy 169).

Although Yeats defends Synge's plays and uses his greater fame to bolster Synge's career, their style and temperament differ greatly. Yeats' attraction to the medieval past is aristocratic; like T. S. Eliot, he believes that the hierarchical order of class is a positive and "natural" state. Synge, on the other hand, identifies with society's outsiders: tinkers, travelers, peasants, and playboys (used here, the word means *sportsman*, not *philanderer*). Only in *Riders to the Sea*, among his plays, does a positive description of crowd or community emerge. Otherwise, his heroes are marginal figures who stand apart from community, in the Romantic tradition, or are exiled by it, because their integrity threatens and is threatened by the hypocrisy and vices of the mainstream (McCarthy 167–8).

Many contemporary scholars of the Irish Literary Revival attribute Synge's choice of Aran to Yeats' peremptory direction in 1896: "Give up Paris; you will never create anything by reading Racine, and Arthur Symons will always be a better critic of French literature. Go to the Aran Islands. Live there as if you were one of the people themselves; express a life that has never found expression" (Price 78). The agency of Synge's decision is now in dispute. Declan Kiberd believes, for instance, that Synge's acquaintance with the Sorbonne's Professor de Jubainville, and his book, *The Irish Mythological Cycle and Celtic Mythology* (1904), deepened Synge's comparative literary tendencies and re-directed his attention from Continental writers to the language and lore of his own Ireland. As Kiberd puts it, Yeats was an important factor, but Synge "was heading in that direction from the very beginning" (*Synge and the Irish Language* 37). Synge and Yeats were their own artists, though collaborating in a movement much larger than themselves. Synge, Yeats and Lady Gregory formed a good directorship of the theater, in part because they adopted such different approaches to representing (or inventing) Ireland.

John Millington Synge in July 1906. A photograph taken in Edinburgh during the Abbey Theatre Company visit. Ann Saddlemyer describes her impression of Synge, "The nervous and virile face was pale beneath his summer tan, the cheeks rather drawn through constant illness; it was not a handsome face, but singularly expressive and sometimes almost sad. His light hazel eyes, at once smoky and kindling, gazed directly and frankly" (*Letters* ix). From *J. M. Synge and His World* by Robin Skelton, Thames & Hudson Ltd., London.

"Virile Virgin"[10]: Women in Synge's Life

Having claimed that the Aran experience made Synge into a man confident enough to court an attractive, spirited, seventeen-year-old actress, I should explain his history with women prior to casting Molly Allgood as a walk-on in *Well of the Saints* in 1905. Synge's was a fairly typical love history for a good-looking, sensitive, agnostic writer of his time—full of false starts and rejected proposals, but culminating in a relationship that crossed barriers of age, class, religion, and temperament. His first love was cousin Florence Ross, who lived next door to the Synges in their suburban home outside Dublin (in Orwell

Park, Rathgar). At ages eleven and twelve, the children shared an interest in nature. They wandered the woods, discussed flora and fauna, and co-authored a naturalist journal. Synge loved her, and remembers kissing the chair that she had vacated. In retrospect, he regards the interlude as a sort of lost paradise, the happiest time of his life (Kiely 52). He and his cousin were natural with each other, unworried about the sexual implications of their friendship. They knew all about the birds and the bees, and talked with "frank amusement" about sex, much like the peasants whose unselfconsciousness he later admired. Eventually their interests diverged, Florence taking up water-color painting and John, though remaining an amateur naturalist, switching from birds to insects. They drifted apart, as children do, without any apparent trauma to John.

The next relationship was similarly convenient, as Cherrie Matheson's family moved three doors down from the Synges' summer-lease in Greystones, a popular Protestant enclave. Cherrie's father was a leader of the Plymouth Brethren, a strict, fundamentalist sect with Calvinist ideas about salvation. Cherrie believed, in the fervency of her twenty-one years, that she could convert or remake John into a believer. For his part, John was open about his agnosticism, enjoyed debate with an intelligent woman, and did not think that this obstacle should defeat their love. He carried the torch for her throughout the exciting travels of his youth — six months in Germany studying music, and five in Paris studying languages. Proposing once by mail and a second time in person, he was twice rejected because of their religious differences. Perhaps because it was his first serious relationship, perhaps because of the connections to his family, perhaps because she was highly intelligent, but even more likely because she rejected him, Cherrie Matheson played the mythopoeic function, in John's life, of Heartbreaker. Their frustrated courtship inspired bad poetry, melancholy moods, and low self-esteem. Fortunately, he simultaneously developed a series of friendships with other women, who gradually built his self-esteem. These warm friendships established a pattern in John's life: women rather than men were his confidantes, whose thoughts and feelings he respected (McCormack 36). In his plays, women hold interesting and central roles; they speak powerfully, especially for women of their time.

If Cherrie Matheson plays the role of pious heartbreaker (The Holy One, as John's friend Valeska von Eiken nicknamed her), then Molly Allgood might be Paradise-Nearly-Regained. At the time of his death in 1909, she had assented to his marriage proposal, they were both in love, and his mother tacitly assented to a match that offended all of her instincts of decorum. Ironically, Mrs. Synge did not warm to Molly until they had shared a watch at John's sick bed. Marriage was prevented only by John's death from Hodgkin's disease. Molly had attracted Synge's attention in her walk-on part in *The Well of the Saints*. She was pretty in a particularly Irish way; as Synge's mother describes it, she had a combination of virginal Irish innocence and "strong 'come-hither' well calculated to drive men mad" (McCormack 288). She was also a good actress, appropriate for Synge's parts, with her vivacity, her rich voice, and her working-class manner of speech. He decided to cast her as Cathleen in *Riders to the Sea*, a practical, stoic girl in the poverty- and grief-stricken Aran Islands. Impressed by her handling of the role, he next read her for the lead in *The Shadow of the Glen*. Nora Burke is a full-grown woman who dares express her loneliness and dissatisfaction with a miserly, cold, and much older husband. Not only does she talk to every passing tramp and traveler (as Molly flirts with her fellow actors), she also has the audacity to leave her home with a tramp when her husband catches her in a flirtation with another man. Synge assumes the role of the Tramp in order to coach Molly; the insinuation of love and attraction in the lines

Molly Allgood, Synge's fiancée at the time of his death. She had a charisma equal to Synge's. His mother described it as "a combination of virginal Irish innocence and strong 'come-hither' well calculated to drive men mad" (McCormack 288). From *J. M. Synge and His World* by Robin Skelton, Thames & Hudson Ltd., London. Photograph in The British Museum.

then manifests as real chemistry between the actors (Kiely 158). Because both Lady Gregory and Synge wished to guard the professional propriety of their theater, John and Molly attempted to keep their relationship secret. Their courtship took place during long walks in County Wicklow, in the beauty of the glens that had given his passion voice in *The Shadow of the Glen*.

The May-to-September relationship had its share of difficulties. In Kiely's opinion, it was ironic that Synge should try to create a Molly-Pygmalion along the lines of Henry Higgins correcting Eliza Doolittle, since the freshness and simplicity of her class were the very attributes that had attracted him to her in the first place (159). In my view, the urge to polish the diamond in the rough is only natural in one whose whole ambition in life is to make poetry out of plain folk. He assigned her reading lists full of Shakespeare and the classics, taught her botany to enhance their walks, corrected her pronunciation, and was jealous of her relations with other actors. In turn, she loved him, quarreled with him, was grateful for his assistance of her career, and waited for marriage.

How did the trips to Aran give Synge the confidence to pursue a relationship so ill-regarded by his mother, so unlike any he had attempted before, and so fruitful to his art? They showed him that women did not have to be of his class or religious background in order for him to love them. The Aran Islands opened his eyes to a new kind of female beauty. The women of Aran had a kind of accessibility or naturalness that he had not previously witnessed. It was not sexual, but neither was it sexless. Once when Synge was showing photographs he had taken of the islanders, a lovely girl leaned guilelessly across his knee, and he wrote, "the complete absence of shyness or self-consciousness in most of these people gives them a peculiar charm" (*The Aran Islands* 60). Later, he compares Aran women to those of Paris and New York, in being "before conventionality" and having "liberal features" (95). John is attracted to a young Aran woman (or a few; it is difficult to distinguish them as he does not name them), but perhaps he is rebuffed by her words, "I think it's to hell you'll be going by and by" (68). Possibly she is like the band of women on the docks who turn on John as soon as their men leave the island, "jeering and shrieking at [him] because [he is] not married" (90). Could their animosity be caused by his annoying unavailability, given that he is a scholar with no intentions of settling down on the islands? Another girl (or it may be the same one) requests that he go off-island and get married to a rich woman, then if she dies, return and marry herself (96). Would it be her age that prevents her from indulging the fantasy now instead of later? Apparently, John is attracted to, and attracts, very young women, as is the case with his eventual fiancée. The realization that he is attractive to them is a real gift of the islands.

Finally, there is an artistic process at work in John's interactions with the island girls. Blending the looks on the faces of beautiful island women together with his artistic ideas makes possible, not only the plays, but also Molly in the plays. Synge describes the archetypal quality of one young woman: "at one moment she is a simple peasant, at another she seems to be looking out at the world with a sense of prehistoric disillusion and to sum up in the expression of her grey-blue eyes the whole external despondency of the clouds and sea" (*The Aran Islands* 67). The eyes truly are the mirrors of the soul, for this poetic sensibility. This unnamed Aran girl, or her essence, is Maurya of *Riders to the Sea*, Nora Burke of *The Shadow of the Glen*, and Pegeen Mike of *Playboy of the Western World*, all in one. She is an amalgamation of Irish womanhood, and Synge will make Molly the repository of that ideal. It is not just any Irish woman that Synge wishes to symbolize, however; it is an island ideal, with a "beautiful oval face with the singularly spiritual expression" that suggests a "link between the wild mythology that is accepted on the islands and the strange beauty of the women" (10). Believing in fairies does wonders for the complexion and gives a soulful look to the eyes.

The more academic of biographers can hardly avoid psychoanalysis of their subjects, and W. J. McCormack is no exception. He believes that Synge was traumatized by his upbringing and stifled his own sexuality. The sexual undercurrents of the plays, then, are compensation for his repression. McCormack thinks that Synge chooses Molly because of her adolescent virginity. Molly was "just the girl for Synge because she functioned to advance that abnegation which had its double-origins with Florence/Darwin, which developed a ritual and a pattern in hopeless love for Cherrie, and which found a local habitation and a name on the Aran Islands" (McCormack 428). The biographer is struck by the fact that, while clearly sexually attracted to Aran women, Synge left it at "generalised implications" (427). McCormack believes that, despite his rejection of Protestant morality, Synge remained "powerfully inhibited" about having sex. Despite the gossip that Synge visited prostitutes in Paris, McCormack surmises that Synge died a virgin (426–430). Molly's innocence reassuringly recalled Florence Ross, and was healthier than Cherrie Matheson's "bourgeois conventionalism," while beautiful Aran Island women were merely "mythopoeically fertile" to Synge's imagination, in McCormack's view (428). On the other hand, biographer David Kiely gives the romance far more credit, and cites the love language of *Playboy* and *Deirdre of the Sorrows* as evidence of John's experience in sexual love. "John and Molly are passionate lovers," writes Kiely, and quotes the Deirdre character, "It's a long time we've had, pressing the lips together, going up and down, resting in our arms" (*The Playboy and Other Plays* 200).

Structure and Themes of The Aran Islands

Synge visited the Aran Islands five times, once every summer from 1898 to 1902. The longest visit lasted six weeks; the others were each roughly one month long. His usual home during these visits was the cottage of the McDonoughs on Inishmaan. The middle island is just three and a half miles across, and was then the most primitive of the three — Gaelic was still spoken. Patrick McDonough was happy to host Synge, as he had other scholars of Gaelic; his seventeen-year-old son Martin gave lessons. Synge's bedroom off the kitchen had a wooden floor and ceiling, while the rest of the family slept in two smaller, earthen-floored rooms on the other side of the kitchen. When turf ran low, cow dung had to be burned for heating and cooking, and Synge was driven out by the stench. He would ruminate on the nearby cliffs, which was fine in good weather.

Synge's primary relationships during these visits are with Martin McDonough, Pat Dirane and Old Mourteen (both storytellers) and a few unnamed young women. Pat Dirane figures prominently in the book because he is a storyteller in the ancient bardic tradition. He differentiates between anecdote and folk tale, often establishing authenticity by stressing that he either saw the events of the story or was told these events by a direct participant. Synge's method in *The Aran Islands* is to alternate between documentary-style reporting of events of the day (like thatching, meals, fishing, funerals) and folk and fairy tales, all interwoven with a small amount of his emotional response to the island. His courtesy as a visitor who is also a cultural outsider is demonstrated by many small acts of kindness. In the Introduction to *Aran Islands*, he expresses gratitude to the islanders, and explains the use of pseudonyms as an attempt to protect their feelings, even though he has said nothing bad about them. He converses easily with everyone around him, regardless of age, status, or personality. He carries on a friendly correspondence with Martin during intervals when one or the other is away. Synge follows islanders' ways of doing and never tries to impose his own. He accepts raucous teasing by women, trusts his crew when they take him out in bad weather, and keeps an appointment with a repellant man for the sake of courtesy. Synge was something more than an ethnographer; he deeply respected the islanders as people, not subjects.

His book's blend of emotional response and objective reporting made it an anomaly to the publishing world. Unclassifiable, it was neither the popular exoticism of Pierre Loti (who sold well) nor the armchair anthropology of James Frazer (who earned great respect for *The Golden Bough*), nor yet again the nationalist myth-making of his closest colleagues in the Revival, Lady Gregory and William Yeats. For reasons quite different from the political controversy surrounding *The Shadow of the Glen* and *The Playboy of the Western World*, *The Aran Islands* could not find a publisher until 1907, four years after he had completed the manuscript. The very structure of the book leaves textual questions, prominent among them the nature of the personal changes its protagonist-narrator underwent. Its four-part division makes obvious chronological sense, as each part covers one visit. However, its final twenty pages of fairy stories tend to confuse the issue of personal growth. They are tacked on as afterthought, to appease Lady Gregory, who advised that more folklore and less personal matter should be included. The last sentence ("The next day I left with the steamer") could suggest closure if preceded by some type of lesson or compelling dramatic scene, but instead there are only the fairy stories. Reading Synge's biography does not provide closure on his island experience either, for he returns the next year (that steamer trip was not his last), and tries to return the following year, but is prevented by an outbreak of smallpox on the islands.

Perhaps the structure of the book is better described as cyclical. Each part contains, though it does not revolve around, a major event for the islanders, followed or interwoven with, Synge's reaction to that event. In Part One, the event is eviction. A posse of sheriff's men arrives from the mainland to secure the livestock of a woman who has failed to pay her rent. They remove her from her home and block her doorway with stone. Synge never mentions whether the islanders simply repossess their homes when this occurs. It seems possible, given how difficult it would be to enforce the law from the distant point of the mainland. Synge's response to the eviction is perhaps not quite as emotional as it is literary, though he clearly takes the islanders' side. In typically ironic fashion he observes the approaching posse: "after my weeks spent among primitive men this glimpse of the newer types of humanity was not reassuring" (*The Aran Islands* 44). The literary value of this horrific proceeding is the impassioned speech of the turncoat's mother. For years, no one would betray the islanders' solidarity by identifying the livestock of the tenant-in-arrears to the British constabulary, but now a man named Patrick has done so. In response, his mother waves her arms with "extraordinary rage" and cries, "This man is my own son; it is I that ought to know him. He is the first ruffian in the whole big world" (46). Such expression closely resembles the closing line of *The Playboy of the Western World*, when Pegeen Mike laments, "I've lost the only playboy of the western world"(146). More significant than the similarity of phrasing is the similarity of the instinct to protect the criminal, which Synge notes "is universal in the west" (*The Aran Islands* 50). Hence, the mother castigates the son who turns in the criminal; the son becomes an outcast by aiding the Law. Synge's best-known play centers on this irony; *Playboy of the Western World* is about a man who has purportedly slain his father, and is admired for it and granted protection from the law.

The major event in Part Two is Synge's stormy passage to Inishmor, the big island. The difficulty of the passage gives him immense respect for both the sea (a "world of inarticulate power"), and the skill and courage of the rowers who face it (*The Aran Islands* 63). This section is characterized by melancholy moods brought on by dark weather. The island is quite different in September than in May. Synge's initial experience of the rain, cold, and gray skies is overwhelming: "after a few hours the mind grows bewildered with the endless change and struggle of the sea, and an utter despondency replaces the first moment of exhilaration" (63). To increase the sense of doom, Martin's mother spends an evening keening the loss of her sons. Both are alive, but since they are off island, it seems they are lost forever. It is not the first time Synge has heard keening[11] (there was a funeral in Part One), and it is not the last (there is another funeral in Part Four), but it is the time that gives him the deepest understanding of motherhood. Maurya is the old woman protagonist in *Riders to the Sea,* who loses all six of her men to the sea. Martin's mother inspires this fictional character. Synge sums up her life and the play's theme in these words: "The maternal feeling is so powerful on these islands that it gives a life of torment to the women. Their sons grow up to be banished as soon as they are of age, or to live here in continual danger on the sea; their daughters go away also, or are worn out in their youth with bearing children that grow up to harass them in their own turn a little later" (62). Have a little biological determinism with your environmental difficulties!

Part Three's major events are the drowning of men on the little and middle islands, as well as Synge's trip to Inisheer, the little island. On that trip, Synge's encounter with a rogue evokes the deepest response. This is the man for whom Synge puts on a face of liberal cordiality, even though he feels repelled. His "inexplicable loathing" recalls another primitivist, D. H. Lawrence, and his strong reaction to a similar laborer named Il Duro ("The Hard") in *Twilight in Italy.* Il Duro is a vine grafter working on a farm in northern Italy where the Lawrences are staying. The worker repels Lawrence because of his cynicism towards women and his consequent aloneness. Synge analyzes his rogue in a similar way, describing his look as "a curiously mixed expression of hardness and melancholy," and supposing that his character has given him a bad reputation so that he has no companions (*The Aran Islands* 94). In both writers, an uncanny mirroring effect takes place; there is something they identify with but do not want to acknowledge. Synge is quite alone on the islands, in the sense that he is always the outsider there. Lawrence, though in a committed relationship with Frieda at the time of the Italian trip, fears what could have been if he had not found her, and what might still be if she returns to her children. Perhaps for Synge, the hard laborer serves as the symbol of his own difference from islanders, even while the man's isolation sparks a frightening sense of recognition.

Part Four's major event is the funeral and burial of a young man who drowned. The keening sounds, as well as strong visual components, give Synge materials for his work. In the family plot, skeletons must be unearthed in order to fit the new corpse into the small space. During this excavation, the mother of the dead man finds and cradles the skull of her own dead mother, keening wildly. The gesture recalls Hamlet holding up the skull of his erstwhile friend Yoric, but the tone is totally different. While Hamlet orates philosophically about the leveling power of death, the Aran woman is wordless, though "keening and shrieking over it with the wildest lamentation" (*The Aran Islands* 113). The other women also beat upon the coffin and keen till they are prostrate, while the men's voices crack with the strain of the continuing keen. Synge appreciated the powerful emotive value of the keen. So convincingly did his actors keen in *Riders to the Sea* that, after a London performance, audience members went away "keening down the street," according to Lady Gregory (Kiely 129). *The Aran Islands* should have ended with this funeral. Instead, there are twenty-two more pages of fairy stories. By refusing to end his book in his own voice, Synge was responding to both Lady Gregory and Yeats's criticisms that the book was too personal. The sad effect of his compliance to their suggestions is that Part Four, and thus the book, lack closure. Fortunately, Synge later learned to trust his own judgment.

Bringing the Moods of Aran to Fruition

After Aran, Synge wrote six excellent plays in succession, two of which (*Riders to the Sea* and *Playboy of the Western World*) are considered masterpieces and frequently anthologized in drama texts. He also penned *The Aran Islands*, which, if not a masterpiece, is an original work in its own category, a book in which he found his themes and voice for future writing. This compassionate though factual voice also informs his other prose work of the period, essays on the spirit of place called "In Wicklow," "In West Kerry," and "In Connemara." The trip to Connemara, a desolate county just north of Galway, and Ireland's poorest, was commissioned by England's newspaper *The Manchester Guardian.* The assignment was to report on the results of the Congested Districts Board's efforts to ameliorate poverty in west Ireland. The interesting misnomer "congested" refers, not to a population explosion, but to the fact that the lands were too infertile and too closely partitioned for natives to eke out a living. Although purportedly objective, the twelve Connemara pieces manage to convey Synge's attitudes about many subjects, including the degradation of peasant life by poverty, the questionable efficiency of the CDB's interventions, and the perfidy of shopkeepers who take advantage of their power as sole middlemen in isolated areas. Both the details and the tone sustain interest at a far higher level than most journalism.

The quality of the six plays, several essays and *The Aran Islands* contrasts markedly to Synge's pre–Aran work. Before Aran, Synge had written sentimental poetry, an aborted autobiography, an aborted novel, and three or four very bad plays. The *fact* of Aran's influence on Synge's creative genius needs no belaboring, as the chronology and improved quality of his production after the experience speak for themselves. The *nature* of that wide-ranging influence bears examination. Aran helped define every literary component of the work — its plots, language, themes, and tone. I will describe Aran's influence on these components one by one.

The plots of four of the six plays were given on Aran. In the first days of his sojourn, the oldest man on Inishmaan tells the story of the Connaught son who killed his father with a spade, then fled to the island to be hidden by some relatives in a hole until it was safe to proceed to America. This in a nutshell is *Playboy*. There is a more sordid source as well, which Synge admits but does not publicize. Well known to the Irish, British, and eventually American public, the James Lynchehaun case was eventually puffed into propaganda for Irish nationalists. In the 1880s, Lynchehaun was groom boy for a beautiful, rich, British lady on Achill Island, County Mayo. When she rebuffed his amorous advances, he took revenge by burning her barn and attacking her person. He bit off her nose, struck her in the head with a blunt object, and kicked her in the vagina. He was sentenced to life for attempted murder, but he escaped to America. Britain tried to extradite him, but the U.S. Supreme Court held it unlawful. Through twists of rumor and propaganda, James Lynchehaun came to stand for the *innocent* Irish patriot, hero of the common man, the very type of the Irish peasant abused by British aristocracy. If Synge were thinking about Lynchehaun when developing Christy Mahon, it would certainly have added color to his portrait, yet he seems to have been more interested in the villagers' response to the hero than Christy's inherent aggression, and Christy has a sweeter way with women than Mr. Lynchehaun.

Around the same time as Synge hears about the fugitive father-slayer, Pat Dirane tells the story of *The Shadow of the Glen*. Once, when Pat was walking from Galway to Dublin, he sought shelter in a roofless house. The lady of the house asked him to watch over her husband's corpse while she went in the next room with her boyfriend. The "dead" husband arose, went into the bedroom, and bloodied the lover with a stick. This is the kernel of *The Shadow of the Glen*.

Another islander tells the story of a Connaught woman who dreamt of a healing well on an island. When learning that such a well existed on Inishmaan, she brought her blind son to it. Sprinkled with its holy waters, he regained his vision. Such a powerful cure is the basis for *The Well of the Saints*.

The play with the most grounding in observed events and reported facts is *Riders to the Sea*. We recall that Synge has attended two island funerals and witnessed Mrs. McDonagh's keening for sons who are not yet dead. In *Riders to the Sea*, a mother has already lost five of her men at curtain's rise. During the course of the one-act play, she loses her sixth and final male relative. Details about her few possessions, how they are used, the folklore she believes in, the negligible role of the priest, and how she copes with tragedy are all directly present in *The Aran Islands* as real things that Synge saw and heard.

Besides the plots, the language of the plays was also given by Aran. In his direct reporting of speech in *The Aran Islands*, and in letters written to him by his friend and tutor Martin McDonagh, we find many of the lines, both grand and small, that are present in the plays. Other than these direct borrowings, we also know something about his method of creating or recreating dialect, because he occasionally defines its components. In *The Aran Islands* he describes the "exquisite intonation" of the earlier-mentioned humorous girl who wants to marry him: she "broods and coos over every syllable" (95). Synge had to teach his actresses how to coo their speeches, and Molly was a quick study. The Aran girl also "plays continual tricks with her Gaelic in the way girls are fond of, piling up diminutives and repeating adjectives with a humorous scorn of syntax" (96). Nora Burke of *The Shadow of the Glen* does the same: "Isn't it a long while I am sitting here in the winter, and the summer, and the fine spring, with the young growing behind me and the old passing, saying to myself one time, to look on Mary Brien who wasn't that height, and I a fine girl growing up, and there she is now with two children, and another coming on her in three months or four..." (22). This long, incomplete sentence filled with "ands" is one we can easily imagine delivered in the "inarticulate chant" adopted by the Aran spinner (96).

The foremost scholar of Synge's language, Declan Kiberd, notes other values of the Irish language that Synge carries over to the Hiberno-English hybrid, particularly its "hardness" (concrete, rather than abstract words), "easy comprehensibility," and "terse rhythms" (*Synge and the Irish Language* 72, 77). People in Synge plays speak in blank verse, without appearing incongruously Shakespearean. Christy courts Pegeen with such language: "It's little you'll think if my love's a poacher's or an earl's itself when you'll feel my two hands stretched around you, and I squeezing kisses on your puckered lips till I'd feel a kind of pity for the Lord God is all ages sitting lonesome in his golden chair" (*The Playboy of the Western World* 136). In addition to the lilt of the predominant iambs, the occasional three-beat stresses (*two hands stretched*), spondees (*Lord God*), and trochees (*puckered, lonesome, golden*) provide the characteristic emphasis and variation of Synge-song.

Synge clearly loved the Aran dialect, but what made it better than that of Galway, the Blaskets, Connemara, Wicklow, or Dublin? The unique appeal of Aran-talk is evident from the opening pages of *The Aran Islands*. The difference in intonation strikes Synge forcibly. On the first page, meeting "a band of tall girls," Synge focuses immediately upon their speech: "they called out to me with humorous wonder, speaking English with a slight foreign intonation that differed a good deal from the brogue of Galway" (6). The distinctive sound lies in the humor, vitality and purity of their speech, as "they hurried past me with eager laughter, and great talking in Gaelic" (6). Later, Old Mourteen, the blind storyteller, speaks with such "exquisite purity of intonation" that it makes Synge cry (10). For a detached and reserved man, this is no easy feat. Mrs. McDonough, his hostess on Inishmaan, evokes a similar pathos while reciting *The Love Songs of Connaught*: "she recited the verses with exquisite musical intonation, putting a wistfulness and passion into her voice that seemed to give it all the cadences that are sought in the profoundest poetry" (65). Even the little girls of the island "speak with an exotic intonation full of charm" (9).

The sound and musicality of the speech, then, impact Synge viscerally. But the constructions and syntax make a deep impression as well. Speaking with a group of men, Synge differentiates between those who have "a curiously full vocabulary" in English, and those who "are driven to ingenious devices to express their meaning" (15). Ingenious devices are a feature of Aran speakers in full command of English as well, if Synge's characters' speech is any indication. Pegeen mocks Shawn, her wimpy suitor, with ingenious devices: "And you think you're a likely beau to go straying along with, the shiny Sundays of the opening year, when it's sooner on a bullock's liver you'd put a poor girl thinking than on the lily or the rose" (*Playboy* 139). It is not that her comparison of a wimp to a liver is that striking; it is the extravagant way she winds her syntax around the central image that makes her speech noteworthy.

Although Synge reports many conversations with the folk of the Blaskets, Wicklow, and Connemara, he did not fall in love with their dialect. In his essays on other places, he does not describe intonations or syntax of their dialects, but merely tells the content of what is said. Perhaps this is because their speech is rougher than that of the Aran Islanders, in concordance with their personalities. Although Synge did not (and out of courtesy would not) state that the Aran Islanders were the gentlest of all the west Irish people he met, such a comparison can be implied by the respective depictions in *The Aran Islands*, "In Wicklow," "In West Kerry," and "In the Congested Districts." Nor is the gentleness to be taken simplistically, for it includes wildness and passion. The quality of gentleness peculiar to these islanders is one of peace and dignity: "This little corner on the face of the world, and the people who live in it, have a peace and dignity from which we are shut for ever" (*The Aran Islands* 114). Synge suggests that isolation has made it possible for the islanders to retain some qualities of their most ancient ancestors, what he calls an aristocratic primitiveness: "Their way of life has never been acted on by anything much more artificial than the nests and burrows of the creatures that live round them, and they seem ... to approach more nearly to the finer types of our aristocracies ... than to the labourer or citizen, as the wild horse resembles the thoroughbred rather than the hack or cart-horse" (*The Aran Islands* 21).

The islands' separation from the rest of the world gave its people sing-song language and quaint manners, but how did Synge mold Aran language and plots into themes of universal value? Synge names his main theme when he writes, "I cannot say it too often: the supreme interest of the island lies in the strong concord that exists between the people and the impersonal limited but powerful impulses of the nature that is around them" (Kiely 115). Like Gauguin before him, like Lawrence after him, and like other primitivists of the modern era, Synge is most impressed by a population which respects nature and concedes its mastery rather than trying to impose a human will upon it. This concordance is artistically expressed when the people seem to *be* their environment, to blend into it as parts of a greater whole. Noticing the oarsmen's faces when traveling in a curagh, Synge remarks: "Even after the people of the south island, these men of the

Inishmaan seemed to be moved by strange archaic sympathies with the world. Their mood accorded itself with wonderful fineness to the suggestions of the day, and their ancient Gaelic seemed so full of divine simplicity that I would have liked to turn the prow to the west and row with them for ever" (*The Aran Islands* 94). This is not to say that they do not feel the tragedy of their risky lives. In Synge's opinion, keening expresses their rage towards the most basic fact of their existence. Living on small islands in a wild sea, they are always vulnerable: "In this cry of pain [the keen] the inner consciousness of the people seems to lay itself bare for an instant, and to reveal the mood of beings who feel their isolation in the face of a universe that wars on them with winds and seas" (*The Aran Islands* 31).

Although full of pain, keening brings release. After cries of despair, there is acceptance. There is no alternative, and their acceptance of that reality is what makes the people both simple and great. This acquiescence to fate is most directly stated in the tragedy *Riders to the Sea* when Maurya is confronted with the final death of the last man of her family: "Michael has a clean burial in the far north, by the grace of the Almighty God. Bartley will have a fine coffin out of the white boards, and a deep grave surely.... What more can we want than that? ... No man at all can be living for ever, and we must be satisfied" (12). Cutting her losses, Maurya looks at the bright side of death: "It's a great rest I'll have now, and great sleeping in the long nights after Samhain [because she won't have to worry about her men], if it's only a bit of wet flour we do have to eat, and maybe a fish that would be stinking" (11). The loss of the men means less food for the household, but the trade off—fewer worries— might be worth something after all. If it were not tragic, it would be comic (the truism that women are always worried about material goods), but the combination of the tragic and comic modes makes it ironic. Synge brings out the ironies in Aran ways of saying and doing, without ever diminishing them. Theme becomes indissoluble from tone when the attitude of acceptance is expressed in an intelligent literary way, through irony.

One of my favorite ironies in *The Aran Islands* is an old man's statement that is intended to assuage Synge's fears about embarking on a "gulf of struggling waves" in a small curagh: "A man who is not afraid of the sea will soon be drownded, for he will be going out on a day he shouldn't. But we do be afraid of the sea, and we do only be drownded now and again" (70). This is wry humor, if only to the outsider. First, the outlook that losing only *some* people is going to cheer Synge seems absurd, since he, the foreigner, could easily be one of the unlucky ones. Second, the situational irony abounds in the fact that the squall is as black and nasty as any Synge has ever seen, yet the islander insists that this is not one of the bad days.

This type of gallows or black humor, which is apparently not humorous to the speakers themselves, appears frequently in *Playboy of the Western World*, a comedy. When Pegeen's father leaves his daughter alone with Christy Mahon to protect her from tinkers and peelers (police) in the dark night, one of his drinking companions applauds the choice: "Now herself be safe this night, with a man killed his father holding danger from the door" (107). It is hard to believe that Jimmy can deliver this line with a straight face, but the audience needs to suspend its disbelief and buy the premise of this play, that the Mayoites respect a man who kills out of passion, because it means his instincts are pure. Apparently, they suspect rape only from tinkers, not from an unlikely lad running away from a potato farm.

If we are not sure of the playwright's ironic intention in the previous instance, we are sure when he states it in this description of the grievous funeral in Part Four of *The Aran Islands*. After keening, there is a great falling-off of intensity, as the (much smaller) Christian portion of their ritual is satisfied: "Before they covered the coffin an old man kneeled down by the grave and repeated a simple prayer for the dead. There was an irony in these words of atonement and Catholic belief spoken by voices that were still hoarse with the cries of pagan desperation" (*The Aran Islands* 32). No more succinct expression of the relative wholeness of pagan practice can be found in these pages. Raging against the indifference of nature is a necessary part in the grieving process; acquiescence that it is part of God's greater plan comes later, and in a much quieter way. In keening, the peasants seem to blend in with their environment of the sometimes-screaming wind and waves; in Catholic prayer, their ritual seems extraneous, inorganic, superimposed. Maurya expresses this ironic discrepancy when Nora tries to assuage her fears for Bartley by repeating the priest's words, "the Almighty God won't leave her destitute with no son living." Maurya responds, "It's little the like of him knows of the sea" (*Playboy and Other Plays* 9). By sea, Maurya means more than just meteorology and nautical wisdom; she means the god of the sea, the Celtic Manannan Mac Lir, who claims lives on a regular basis. While the priest banks on a merciful God, Maurya knows an indifferent one, the "impersonal ... powerful forces of nature" that Synge sees in Aran. In this tale, the priest is wrong, and Maurya is right. Pagan interpretation of life prevails.

Not all of the ironies that Synge observed on Aran are so bleak, yet they are all subversive, which is what makes his plays both controversial to his contemporary audience, and appealing to moderns. All of the plays contain structural and situational ironies. In Aristotelian terms, the *peripeteia*, or sudden reversal in events, is essential to both tragedy and comedy. Synge excels at truly unexpected *peripeteiae.* In *Riders to the Sea*, the *peripeteia* is least obvious, since death is a sure thing from the very beginning. However, the mode of Bartley's death is ironic because it is not caused directly by the sea. In her vision, Maurya sees a resurrected Michael on the pony that Bartley is leading to the quay. Bartley's mode of death seems uncanny, then, when the villagers report that the grey pony knocked him off a cliff into the sea. Did Michael's ghost kill his brother? Fairy stories hold that the dead want their closest companions with them on the other side.

In *Playboy*, Acts II and III contain so many reversals that they oddly justify the final surprise, which is that Christy leaves the first love of his life to tramp with his much-hated father. In abbreviated form, the plot of the play is this: Pegeen Mike is promised to a certain Shawn Keogh, a far too timid man for the strong woman that she is. Enter Christy Mahon into Pegeen's shebeen (drinking premises): weak and small, he is nonetheless attractive, and gains much in stature when he tells the bar patrons of his plight. In a pique of anger he has dealt a blow to his father's head with a spade, then fled to escape repercussions. He makes love to Pegeen, who is ensnared, then proves his manhood by winning the day's sporting competition. Although transported to the bar on the shoulders of the crowd, he soon loses his hero status. Very much alive, his father has wandered into the bar while trying to find his son. Once again, Christy batters Old Man Mahon, in hopes of keeping Pegeen, but the villagers are angry at the lie, and afraid of the violence when it occurs under their noses. They tie him up, Pegeen burns his leg with a turf ember, and they prepare to turn him in. In the nick of time, the father rises again and frees his son, who proclaims that he will roam the countryside, "romancing through his lifetime." His father and former master will now be his slave.

We know that Aran provided the foundational story about villagers protecting a fugitive, but how did Aran influence the additions that make this story great and comical? The love story centers on the power of talk to aggrandize a man in women's eyes. Surely Synge did not witness any courting on the island as elegant as the words of Christy and Pegeen? No, but he did experience the sauciness of girls left to their own devices on the stark roads of Inishmaan: "The women and girls, when they had no men with them, usually tried to make fun with me. 'Is it tired you are, stranger?' said one girl. 'Bedad, it is not, little girl,' I answered in Gaelic, 'it is lonely I am.' 'Here is my little sister, stranger, who will give you her arm.'" (*The Aran Islands* 38) Although Pegeen lets Christy do the courting, another woman (the Widow Quin) aggressively pushes her own case for marriage within the first few minutes of meeting Christy.

Also reminiscent of *Playboy*'s villagers are the Aran women on the quay who turn on their honored guest John as soon as their men are launched: "The women were over-excited, and when I tried to talk to them they crowded round me and began jeering and shrieking at me because I am not married. A dozen screamed at a time, and so rapidly that I could not understand all they were saying, yet I was able to make out that they were taking advantage of the absence of their husbands to give me the full volume of their contempt" (*The Aran Islands* 90). Like the Mayo crowd, these women's reactions to Synge run the gamut from desire to contempt, and like the Mayoites of his play, they act in unison, creating a spectacle. Although bad for his ego, this was very good for Synge's future *Playboy*.

The dramatic ironies of the three other comedies are even more obvious. They demonstrate Mikhail Bakhtin's "carnivalesque" theory[12] which states that lower classes often subvert the upper classes, by exposing the vices of the high while suggesting certain virtues of the low. In *The Tinker's Wedding*, Sarah and Michael are tinkers in a common law marriage who, on a whim, decide to get legally married. When the priest turns on them, they tie him up, but release him on his word not to tell on them. He tells on them only to God, and they run away, apparently scared of his curses. The underlying irony of this plot is that they are better off without the marriage, as it symbolizes the trappings of the establishment. Dramatic irony occurs when the audience knows more than the players. In this case, the audience sees that the tinkers' pagan ways are healthier than the Irish society's, as represented by the vice-ridden priest. As the curtain falls, these tinkers shouldn't remain scared for too long. They have a folksy way of justifying everything, as long as there is a "bite to eat, a drop to drink, and a time of love when they are young and fine" (*Playboy and Other Plays* 50).

The Well of the Saints is based on an ironic structure as well; it proclaims the superiority of the wanderer to the questionable morality of the establishment. Two blind, married beggars regain sight from holy water. Seeing that his wife is ugly, and regretting that with sight comes the necessity to work for a living, the husband decides to remain in darkness once his sight relapses. The wife agrees. The priest and villagers turn against them for this decision, believing there is something subversive about the choice of darkness over light. However, the audience sees that the beggars are nobler because they find joy in the signs of nature that penetrate their blindness. Smells and sounds of the flora and fauna are sufficient for the beggars, whereas the townspeople are consumed by envy and greed, the vices that go with sight.

The Shadow of the Glen also holds the tramp's life higher than the grasping sterility of its two working men. Nora is married to a stingy, cold shepherd, who plays dead in order to surprise her and her lover. Ironically, his punishment spells liberation for her. He turns her out of his house, but the visiting Tramp is happy to show her the joys of the wandering life and save her from the cramped life of the shadowy glen. She had been sitting in a lonely house, deteriorating slowly, waiting for death. In each of the plays, outside is opposed to inside as a far superior place. Nature rather than community provides the fundamentals of life. Community corrupts one's understanding — the message is not only subversive; it is dire. But it should not be taken literally; it is literature. Like a satirist, Synge pokes fun at those institutions Ireland takes quite seriously, marriage and church at the top of the list. Unlike most satirists, he poses an alternative, unlikely and abstract as it may seem. Offended audience members may have thought he was exhorting all people to leave their homes and beg for their bread like tinkers and travelers. That is a ridiculous extrapolation. What his work does suggest is that modern humans live in a wounded state, both psychologically (Julia Kristeva's "original wound" is birth-separation from the mother) and culturally (the alienation and angst of modernity). But, if we are receptive to it, nature can heal what is most wounded in us. Literature plays an important role in this receptivity since it allows us to appreciate how others have negotiated the human condition.

In Aran, Synge found a people living the greatest portion of their days outside. Their struggles were with nature more than with people. This is not to say they did not quarrel — one of the island pastimes was eavesdropping outside one family's window while they traded abuse for four hours at a time. Synge may sometimes idealize his peasants, but he also clearly registers their few weaknesses, such as the failure to show compassion for the pain of animals and humans. They also fail to exercise caution about rowing when drunk. More profoundly, the evils of capitalism are insidiously encroaching . He observes the "anxiety of men eager for gain" — but that is on Inishmore, and clearly distinguishes it from Inishmaan (*The Aran Islands* 69).

Many biographers and critics characterize Sygne's Aran experience as seminal in his life and literature. Battaglia and Roche describe Aran as healing Synge's inner division between contradicting pagan superstitions and Christian orthodoxy. Weldon Thornton believes Aran enabled him to break free from rationalism (Morrissey 21). Alan Price speaks of the microcosm of Aran, which offered all that Synge considered most important in life in a condensed form, accessible to analysis. Richard Fallis' claims are perhaps closest to my own, in that he describes Aran as offering Synge a self: "He has created [in *The*

Aran Islands] a self ready to become the artist known as J. M. Synge" (151). However, I do not agree with Fallis when he says that *The Aran Islands* demonstrates an increasing identification with the Aran people. Fallis writes, "The narrator's growing sense of control of himself is here a central theme, that exemplified by his growing command of the Irish language and a parallel sense of identity with the Aran people" (151). A comparison of Parts IV and II, for instance, does not prove a greater sense of identification; Synge always knows that he is apart, that a great gulf separates him. In Part II, he observes that, "in some ways these men and women seem strangely far away from me" (66). He feels "a waif among the people" when he realizes that he cannot communicate to them his deeper thoughts and feelings: "I can feel more with them than they can feel with me, and while I wander among them, they like me sometimes, and laugh at me sometimes, yet never know what I am doing" (66). He is finding a "self" by defining it both with and against them. Although he does not bring up his own loneliness again in Part IV, this does not mean that it has passed. On the contrary, he has used his own sense of separation to enhance the way he describes it in others. The funeral in Part IV achieves a drama and literariness absent from the earlier funeral description. Death surrounds the mother of the dead man; generations of death hug her like bookends, as she unburies her mother in order to bury her son.

Aran did help Synge find a self, but not through a complete identification with the islanders. Instead, through comparative cultural analysis, Synge accessed the collective unconscious that psychoanalyst Carl Jung would write about within the next decade. The collective unconscious links all peoples, of all times and places, by certain recurring archetypes in their collective imagination. While Jung used the idea for therapy, Synge used it to make his plays about local, rural life resound with universal value. For instance, he immediately recognized the tale of the unfaithful wife told him by Pat Dirane on Aran, as the Widow of Ephesus, one of the most common folktales in western Europe. Consistently, he looked for connections between the islanders and the medieval or ancient history of their forebears. He compared the islanders' gentleness to the "refinement of old societies," reasoning that they were able to retain these fine qualities because the island's isolation preserved them from the tramp of "the heavy boot of Europe" (*The Aran Islands* 21). Even in his own day, he sees a different type of peasant on the islands than the mainland, as well as a different type from island to island. The people of the small island, Inisheer, are a "ragged, humorous type" that used to be the norm of the Irish peasant (again, one that was wiped out by the heavy boot of modernity) (*The Aran Islands* 92).

Aran gave Synge a profounder sense of self because there he learned how to apply the type of comparative cultural analysis modeled by his professor de Jubainville in Paris. The challenge was to blend his scholarly training in the Greek and Roman classics and Celtic folklore with the reality of the rough life and mysticism of Aran's people. He "found himself" in the sense that he did not give up his rationalist, academically trained, Protestant Ascendancy self in order to write the plays about peasants. Like his compatriot James Joyce, Synge blended past and present, local and universal. In *Ulysses*, Joyce humorously related lower-middle-class Dubliner, Leopold Bloom, to the timeless hero-king, Odysseus, whereas Synge subsumed and blended his rich cultural tradition with the traditions of the Aran Islanders. It took a subtle personality to blend these apparent opposites; it is the same sensibility that enabled him to blend his life with theirs as a welcome friend and honored guest.

V

"The Man Who Loved Islands": D. H. Lawrence and His Island Scheme

> For as this appalling ocean surrounds the verdant land, so in the soul of man there lies one insular Tahiti, full of peace and joy, but encompassed by all the horrors of the half known life.—Herman Melville, *Moby Dick*

With D. H. Lawrence, my study turns inward to mental islands as images of withdrawal. After feeling trapped in England during World War I, Lawrence and his wife Frieda finally received government permission to leave in 1919, and were more or less constantly on the move till his death in 1930. Of all the places in which he tried to be happy, his two favorites were New Mexico and Tuscany. Although he often thought about islands, and imagined his personal utopia as taking place on one, he never found an island he liked. His experiences on them, especially the small islands of Capri and Ceylon, proved stifling. Capri was overrun with expatriates: "a gossipy, villa-stricken, two-humped chunk of limestone, a microcosmos that does heaven much credit, but mankind none at all" (KW 556*). Ceylon was literally stifling with heat ("all Ceylon has a lid on it"), and

*Abbreviations used in citing works in this chapter: FU (*Fantasia of the Unconscious*), KW (Kinkead-Weekes, *Triumph to Exile*), LCL (*Lady Chatterley's Lover*), MWLI ("The Man Who Loved Islands"), SS (*Sea and Sardinia*).

imaginatively stifling because he disliked the ideology of passivity in Buddhism (KW 19). His experiences on the big islands Sicily and Sardinia are only slightly happier, but probably in spite of the fact that they are islands. Partly symptomatic of his tuberculosis, Lawrence's restlessness is remarkable—"Comes over one an absolute necessity to move," is the opening to *Sea and Sardinia*. Small places quickly seemed confining to him. *Sea and Sardinia* (1921) records his attempt to find a more primitive way of life on an island known for its outlaws. The Sardinian journey was itself an escape from that other island which he then called home, Sicily, where he felt the locals' backgrounds were too diverse for them to have simple ideas. Sicily had been "subdued" by too many colonizers over the centuries, resulting in a complexity of heritage that troubled him, whereas Sardinia he believed was "outside the circuit of civilisation" (SS 9). To his chagrin, modernity and its problems had already corrupted Sardinia's inhabitants before he arrived.

The stereotypical South Sea Island dream fared no better with Lawrence. After living in Australia for five months in 1922, Lawrence and his wife Frieda shipped to San Francisco, making stops at the Cook Islands and Tahiti on the way. Although he had considered living in the South Pacific, the overnight stay changed his mind. In a postcard to Compton MacKenzie, a sharer of the dream of escape to paradisiacal islands, Lawrence wrote: "If you are thinking of coming here [Tahiti], don't. The people are brown and soft" (iv 286*). With regard to the desire of primitivists to retreat to the innocent state of South Sea natives, Lawrence does not see the appeal: "But we can't go back. Whatever else the South Sea islander is, he is centuries and centuries behind us in the life-struggle, the consciousness-struggle, the struggle of the soul into fullness" (Ellis 56). Lawrence's pull towards islands was different from the other figures in this book. Not aesthetic like Synge's, nor exoticist like Gauguin's, it was metaphorical; islands were a mental space in which to dream utopia, far from the contaminated world of modernity. Lawrence's visionary consciousness was so integral to the modernist literary movement in

*References to *The Letters of D. H. Lawrence* are given by volume, Roman numeral and page number. For example, "(vi 200)" is volume vi, page 200.

England and Europe that to go back to a state of simplicity, even if it were possible, would represent to him a loss in the advanced consciousness that his experimental methods in fiction were striving to attain. So his island had to be a mental construct, with constantly shifting locations, participants, and ideals.

Rananim and Island Schemes

After his experience of Capri, Ceylon, the Cook Islands, and Tahiti, Lawrence's "island" no longer had to be a real island. Over time, he proposed communities of friends in many different places, including Florida and Connecticut, Cornwall and Oxfordshire, the country of Columbia, and the South Seas Islands. Its numbers changed from twenty souls to eight to five to two (himself and his wife). Its cast of characters changed depending on whoever was most sympathetic at the time. Principal disciples shifted from Samuel Koteliansky to John Middleton Murry to Compton MacKenzie to the Honorable Dorothy Brett. The name of the ideal community was Rananim, but after the failure of the first attempt to create it, in 1915, Lawrence referred to it simply as his "Island." The community's ideals metamorphosed from a socialist state without money, to a place to love and create, to a male getaway. After 1926, references to Rananim per se disappear, although he still imagined ideal communion. As Stefania Michelucci notes, it was now a community of two, the man and woman of his ideal union developed in *Lady Chatterley's Lover* (1928). In his very last literary manifestation of utopia, *Etruscan Places* (1932), harmony in community was to be found only in ancient civilizations.[1] In other words, utopia, true to the meaning of its name ("no place"), had finally disappeared off the face of the earth, and was realizable only through imaginative connection with pre–Christian society. The story of his island scheme in perpetual flux is the story of Lawrence's changing philosophies throughout his lifetime.

The history of Lawrence's utopia is compelling, for who among us has not dreamed of the perfect marriage as the base of one's life, enhanced by the perfect circle of like-minded friends to complete one's family-of-choice? Communitarian ideals emerge most strongly in certain turbulent periods of history, such as the 1960s with their hippie communes, the 1920s with their bohemian enclaves, and the 1840s and 1880s with their American religious and secular (and often free love) communities. But the human problem of how to mesh individual happiness with social bonds remains a perennial concern. Living during one of these periods of social upheaval, Lawrence's inherent drive to create a good community, and to define it in distinction from the mainstream, was only intensified by the spirit of the times.

Although Lawrence was depressed many times in his life, the years of World War I were the most excruciating, especially because he could not escape from England, even after being deemed unfit for military service. Being married to a German aristocrat did not help matters; the Lawrences were evicted from their Cornwall cottage on (wrongful) suspicion of spying. They did not receive permission to leave the country until the Treaty of Versailles was signed on June 28, when Frieda returned with trepidation to see her German family in their straitened post-war circumstances, and Lawrence headed for Florence, Picinisco (a small village in the Abruzzi), and then Capri. During the war years, Lawrence's letters show the most frequent references to Rananim. The blessed isle, as he called it, beckoned most sweetly when he was most trapped, indicating a major function of utopia in human psychology. "[Rananim] was a psychological language to focus distress at alienation, and redefine what sense of community was still possible for him to imagine," writes Mark Kinkead-Weekes about Lawrence's earliest references to Rananim in 1915 (182). Of all the sources of alienation during these years, including numerous relocations, frequent breaks with close friends, gathering tensions in his marriage, and money problems, the greatest was the government's destruction and suppression of his novel *The Rainbow* for obscenity in 1915. A lesbian relationship, anti-military sentiment, and sexual feelings were among the suspect descriptions in a novel that seems tame in comparison with the stream-of-consciousness, no-holds-barred writing of the decades immediately following (by Anais Nin, Henry Miller, James Joyce, and Louis-Ferdinand Céline, for instance — but none of these tried to publish in England). England came to represent death for Lawrence, and almost anyplace else offered a greater possibility to host his utopian community than his native soil.

If the choice of place were an essential, albeit shifting factor, then the choice of people was equally important and liable to change. Although Samuel Koteliansky, a Russian exile, writer, and translator, appears to be the one constant on Lawrence's invitation list, he never physically went to the locations of Lawrence's utopia.[2] While many friends talked about the idea, and there were constellations of friends around other Lawrence homes, there are three major constellations of people and places that best illustrate the flux of the idea (and the reality) over time. These are (1) with Bertrand Russell and Ottoline Morrell at Garsington, her country manor, in 1915; (2) with Katherine Mansfield and John Middleton Murry at Zennor, Cornwall, in 1917; and (3) with Dorothy Brett and Mabel Dodge Luhan in Taos in 1922–25.

The first utopian community differed greatly from later conceptions in the sheer grandiosity of its scheme. Starting from a series of public lectures by Bertrand Russell and Lawrence in London, their Rananim was supposed to effect social change throughout England. An ideal intellectual community would spread the good news throughout the nation, until it had become a socialist state. Original participants, during the early months of 1915, included John Middleton Murry, Katherine Mansfield, Gilbert and Mary Cannan, Mark Gertler, and Samuel Koteliansky ("Kot"). When Lawrence met Lady Ottoline Morrell and Lord Bertrand Russell, these two quickly became central to the scheme, not only because they had money, but also because they had that other, less concrete asset: intellectual capital. Perhaps because there was no physical place for these people to live together (though Garsington was a meeting place), there was no actual touchstone of their participation. E. M. Forster and Aldous Huxley indicated their interest at different times; both shared the Modernist viewpoint that the mainstream ethos exerted a deadly influence on the souls of the sensitive. It should be noted that Lawrence did not exclude people from Rananim, but that people tended to drift away on their own accord. The first manifestation of his utopia — Rananim — was therefore little more than a series of discussions among artists and intellectuals.

The first mention of Rananim — its name, motto, badge, flag and description —comes in a letter to Kot (Koteliansky's nickname) dated January 3, 1915: "We are going to found an Order of the Knights of Rananim. The motto is Fier [pride]" (ii 252). Lawrence sketches the badge, a phoenix rising from flames. This image symbolized the capacity of people, individually and collectively, to rise up after trauma (symbolic death) and start life afresh. The phoenix as symbol of hope for renewal continued to appeal to Lawrence for the rest of his life; he would use it on the cover of *Lady Chatterley's Lover* thirteen years later.[3] The name Rananim came from the first line (in Hebrew) of Psalm 33, which Kot used to sing to Lawrence. *Rananim* was a combination of the words "*Rannani*" meaning "rejoice," and "*Ra'annanim*," meaning "green, fresh or flourishing" (KW 182). Both meanings fit Lawrence's hopes for the community, since his vision was essentially religious, conceived in terms of New Testament imagery: "[A man] shall come [to Rananim] because he knows his own soul is not the be-all and the end-all, but that all souls of all things do but compose the body of God" (KW 189). His knights would rejoice in the Lord and flourish in a kind of New Jerusalem. Later, while honing the Rananim philosophy with Bertrand Russell, Lawrence would substitute the Platonic idea of "seeking transcendent truths" for the Christian one of rejoicing in the Lord. During adolescence, Lawrence had rejected Christianity, but he remained deeply religious. He did not mind casting the language of the Bible in a new way to reflect his reverence for cosmic powers, however defined.

When Lawrence met the leading light of the Cambridge intelligentsia in February 1915, the combination of opposites proved both volatile and productive. Bertrand Russell enjoyed the distinction of being the foremost philosopher in England. Lawrence was gradually becoming her greatest novelist. Rationality Personified met Spirituality Personified in a charged historic moment. Both anti-war, these men wanted to reform England radically; she was spiritually and economically impoverished. Their letters over the next seven months display not only the ways in which they modified each other's extremities, but sadly, that their friendship could not withstand their essential differences. In the hopeful springtime, however, Lawrence's Rananim was in full swing, with Lady Ottoline Morrell (Russell's lover) in the center, possibly because she hosted meetings and social gatherings at her country home. Both men agreed that the problem with England (and by extension, Europe) was the "disintegration" brought about by what they called "subjectivism"; under industrial capitalism, where every man is for himself, the notion of community deteriorates (KW 241). They disagreed, however, on the nature of the language in which to formulate the problem and the methods of reconstructing society to cure its ills. The two most fundamental disagreements were about the place of religion (or spirituality) in the vision of the new society, and its political economy. Naively, Lawrence thought he could nationalize industries and introduce equal wages for all. Russell, both more academic and more practical in his approach to social problems, realized how unrealistic this was. In two years' time, the Russian Revolution and Leninist Communism would prove how unsavory such nationalization could be.

Another problem arose with regard to two different perspectives on leadership. Despite his socialist bent, Lawrence believed in a hierarchy of elections: "the electors for the highest places should be the governors of the bigger districts — the whole thing should work upwards, every man voting for that which he more or less understands through contact — no canvassing of mass votes" (KW 248). The belief in a "natural aristocracy" in which better men naturally rise to power, would stay with him for the rest of his life. Russell, on the other hand, believed in the fundamental importance of democratic voting.

Finally, Lawrence believed that the state must be held together by communal philosophical or religious beliefs, whereas Russell believed that "some men can be happy through religion, but there are few" (KW 242). As they tried to collaborate on lecture writing through the summer months, disagreements of both an abstract and a personal nature multiplied. Biographer Kinkead-Weekes thinks that the successful, well-born don, Bertrand Russell, was not used to fierce intellectual opposition, at least not in the direct and personal style of this coal-miner's son. Lawrence's criticisms depressed him, and made him question his faith in his own ideas (KW 243). Lawrence in turn found Russell "immature" for buckling under opposition. On September 14, 1915, Lawrence broke off relations with Russell when the latter refused to support Lawrence's new endeavor, a philosophical literary magazine called *The Signature*, which was his attempt at compromise, an alternative way to present the socialism lectures to the public. Calling Russell a "super war spirit" whose "repressed desires" erupt in "savage jabs," Lawrence asked that they "become strangers again" (KW 263). This marked the ending of the first Rananim.[4] Although Russell and Lawrence were later to reconcile, their collaboration on creating the New World was definitely dead.

Apart from class, political, and religious differences, something more psychologically profound divided them. Mark Kinkead-Weekes sees the root cause of their split in terms of their differing views on isolation versus togetherness. In a letter to Cynthia Asquith, Lawrence declares what Russell "really wants" is "to keep his own little established ego, his finite and ready-defined self intact, free from contact and connection" (KW 258). Lawrence reacted so violently against such islanding or isolation of the self because "he so hated and feared self-enclosure in himself" (258). This diagnosis of Lawrence — that isolation is his deepest fear — is proven brilliantly by the story "The Man Who Loved Islands," which I will return to in the second half of the chapter.

Since socialist revolution proved impossible, the Rananim scheme mutated into what he called the "Island," with a far more domestic outline.[5] No longer believing in the possibility of political reconstruction of England, Lawrence returned to his earlier idea of the Island as escape from diseased society rather than the seed of its renewal. Although he kept on planning Rananim in Florida (because he thought he could find property through Dollie Radford), he also kept postponing the journey until finally it became legally impossible to leave the country because of the draft. His exemption from service was easily enough attained in June of 1916 by attesting to the Examination Board that "doctors said I had had consumption" (even though his biographer does not think Lawrence believed it at the time) (KW 332). The board accepted this diagnosis on the basis of his appearance alone, not requiring any certificate. Indeed, photos of Lawrence during the war years show him beginning to take on that characteristic "lean and hungry look" of the tubercular patient, and of the misunderstood artist.

Renting side-by-side cottages in Zennor, near Land's End, a fitting place for a man chafing to leave England, the Lawrences invited the Murrys to stay next door and form a happy community "in unanimity" (KW 313). John Middleton Murry was well-educated, lower-middle class, and literary — all excellent qualities for friendship with Lawrence. He was no direct literary rival, either, attaining more fame as a literary critic than a novelist. His wife was Katherine Mansfield, a well-born New Zealander whose short stories recall the condensation and irony of Chekhov, while adding the ambiguity and angst of European modernism. The outlook for companionship was auspicious, but any appearance of unanimity or happiness in their Island lasted only about a month, and the Murrys left within about ten weeks. Although Murry was possibly the closest friend Lawrence ever had, and their friendship withstood many breaks, the tensions in the four-way relationship were predictable. For one thing, Murry and Mansfield supported each other's work, and perhaps Frieda was jealous of Mansfield's professional equality with her husband. Frieda fought an ongoing battle against those of Lawrence's friends who disliked her personal strengths (she was outspoken) and demeaned her intelligence. Initially, Murry was one of those; later, when the Lawrences were apart, he would recapitulate and confess his love for Frieda.

As a community of four, though, these strong personalities did not get along. In a witty letter to Kot, Katherine Mansfield describes "all about the COMMUNITY ... Frieda and I do not even speak to each other at present. Lawrence is about one million miles away ... his very voice is faint like a voice coming over a telephone wire. It is all because I cannot stand the situation between the two.... It is degrading — it offends one's soul beyond words" (KW 319). She goes on to describe a horrendous verbal and physical fight between Frieda and Lawrence, ostensibly over the merits of Percy Shelley, which she and John reluctantly witness in their living room.

The reasons for the demise of this second manifestation of the Island scheme are many. First, three of the four members used Ottoline Morrell and Kot as sounding boards, but the sounding boards also gossiped, so that these private grievances were never private.[6] An ensuing sense of betrayal quickly eroded what trust the two couples were able to build by living together. Second, Lawrence's idea of what constituted friendship was very different from Murry's. Lawrence's *Blutbruderschaft*, the ideal of blood brotherhood which his protagonist Birkin proposes to his friend Gerald in *Women in Love*, and which Lawrence proposed to Murry, was in some ways more similar to a marriage (or to his marriage) than a social relationship normally is. In his words, he wanted "no personal obligation, no personal idea. Let it be a union in the unconsciousness, not in the consciousness" (KW 313). But Murry and Katherine were not the "joining kind," as she put it; they based their friendships on intellectual and personal affinity. No wonder she did not like what she perceived as the unsociable behavior of their fellow communitarians.

In addition to the failure of relationship with the Murrys, recurring colds and flu, marital strife, the war, and other broken relations (what Kinkead-Weekes calls the narrowing of the circle) all resulted in a cynical, depressed Lawrence. Even towards unknown villagers in Zennor he harbored distrust quite like the madness he later gives to his character Cathcart in "The Man Who Loved Islands":

> Androphobia. When I see people in the distance, walking along the path through the fields to Zennor, I want to crouch in the bushes and shoot them silently with invisible arrows of death ... the obstructions, the people, like bugs they creep insidiously in, and they are too many to crush.... Oh, if one could but have a great box of insect powder ... [Letters ii 650].

Lawrence uses the term *androphobia* instead of the darker term *misanthropy*, perhaps because he is afraid to admit his feelings have gone that far ... from fear to hatred. The paranoia gets so bad that he begins to think of the Promised Land as just a place, not a group of like-minded people. To Kot in November, he writes: "my Rananim, my Florida idea, was the true one. Only the *people* were wrong. But to go to Rananim *without* the people is right, for me, and ultimately, I hope, for you" (iii 23). To want Kot to go with him gives hope that Lawrence's misanthropy was not absolute; on the other hand, Kot was perhaps the least likely of his friends to leave England and his sheltered life at #5 Acacia Road, St. John's Wood. This idyllic-sounding address takes on symbolic significance because, in contrast to

Lawrence's constant moving, Koteliansky remained fixed at his London address, and that fixedness provided stability for himself and his friend (though Lawrence would also call it "laziness"). Without English citizenship (he was a Russian Jew in exile), perhaps Kot feared to attract the attention of the authorities by applying for visas or moving. In any case, "Rananim without people" suggests that Lawrence is becoming that island of self-enclosure he accused Bertrand Russell of being at the close of their Rananim project. But not for long. One of the attractive things about Lawrence as a man and writer is the certainty of his personal "resurrections"; like the phoenix, he always rises again from the flames of conflict. Time after time, relocation or new people or new books or new offers to publish his own work lifted his spirits and gave him the voice of one reborn.

Although he postponed his voyage to America for another five years after the failure of the Cornwall Island scheme, his time in and around Taos with Mabel Dodge Luhan's coterie between 1922 and 1925 was the closest he came to a true artists' colony, which was at least one aspect of his communitarian ideal. He loved the American southwest as a physical place. It inspired him more deeply than even the Italian countryside he so cherished. The difference between the two was the difference between Old and New World charms: "But the moment I saw the brilliant, proud morning shine high up over the deserts of Santa Fe, something stood still in my soul, and I started to attend.... In the magnificent fierce morning of New Mexico one sprang awake, a new part of the soul woke up suddenly, and the old world gave way to a new" (Ellis 62).

Mabel Dodge Luhan, a rich patron of the arts from Buffalo, New York, moved to Taos in 1917 to establish an artists' colony and to protect and promote local Indian culture. Her interest turned romantic when she married an Indian called Tony Luhan, a local carpenter. She learned of Lawrence from reading *Sea and Sardinia*, though in a bowdlerized version in *The Dial*. Noting both his direct, emotional style and his interest in peasant and primitive cultures, she believed he would suit her purpose perfectly. For such a strong woman, it is startling that her modus operandi with Lawrence — as with other men — was to exert all her considerable power upon him to do her work for her. She was entirely clear that she wished to play Muse to Lawrence's creative spirit: "I wanted Lawrence to understand things for me, to take *my* experience, *my* material, *my* Taos, and to formulate it all into a magnificent creation" (Ellis 59). However, this did not mean submission. As Ellis puts it, "her powerful personality was no more fitted than Frieda's for a minor role amongst the hand-maidens" (59). She gave Lawrence and Frieda a cottage in what he came to call Mabeltown, a collection of small houses around her larger one in Taos. He felt both indebted to her for the financial support, and intensely irritated by her interfering personality. She tried to win him away from Frieda, saying that Frieda was no longer useful for him, and she requested that he collaborate on her own autobiography. This project made Frieda jealous and was soon abandoned.

Meanwhile, Lawrence sought another accommodation to get out of Mabel's immediate range. At the time, Mabel declined to rent them her Lobo Mountain ranch, seventeen miles from Taos, so Lawrence found a wild retreat called Del Monte on his own. Throughout a bitterly cold and snowy winter, they evaded Mabel's immediate sphere of influence, but she and her salon were still the center of arts and letters in Taos. There was no avoiding her if one wanted a bit of social life, and since both Lawrence and Mabel were adept at overlooking past disagreements, their diffused colony survived. At Del Monte, Lawrence and Frieda enjoyed a small-scale version of the Island scheme with two young Danish artists, Knud Merrill and Kai Gotzsche. Lawrence paid the rent on their cabin in return for company (the buffer the marriage needed) and the heavy chores necessary for survival. After the winter, they left for tours of Mexico and Europe before returning to Taos in early 1924.

During his London stay, Lawrence made a concerted effort to attract old friends (the original six of the 1915 Rananim, plus others) to their Taos community. Although Dorothy Brett was the only one to accept, there was another pleasant surprise awaiting them in New Mexico. Mabel offered Frieda the deed to the Lobo ranch, which she had refused to rent or loan the Lawrences the year before. Why did she permit them this retreat from her immediate domain with such generosity, and why was the gift in Frieda's name rather than

The Lawrences at the window of their house in Chapala, Mexico, 1923. This dwelling was the place in which Lawrence completed the first draft of ***The Plumed Serpent*** in two months. Although there was no Rananim here, fellow writers Witter Bynner and Willard (Spud) Johnson lived at a nearby hotel and swam, ate, and traveled with the Lawrences on a regular basis. Photograph Collection. Harry Ransom Humanities Research Center, The University of Texas at Austin.

Lawrence's? Ellis speculates that Mabel may have been concerned about Frieda's complete financial dependence on Lawrence — reminding us that in spite of her imperious style, Mabel did care about others. She also was aware that Lawrence disliked being in her debt (Ellis 182). In any event, the Lawrences did feel beholden, and tried to compensate by giving Mabel the typescript of *Sons and Lovers.* Since it was worth more than the ranch, the sense of indebtedness had merely changed hands. Such considerations of score keeping and payback — so material, so individualistic, and so distrustful — were hardly the basis for the egalitarian community conceived as Rananim. Nor did their subsequent life at Kiowa (as Lawrence christened the ranch) resemble the Island scheme of his dreams. In one particular, however, there was a marked consistency between this and the other Island schemes: all of his invitations save one were refused or ignored.

The Honorable Dorothy Brett, an artist whom the Lawrences had met but not known well during the war years, had been engaged in a love affair with John Middleton Murry immediately following Katherine Mansfield's death by tuberculosis in 1923. When Lawrence returned to England in 1923 (after seeing Frieda off four months earlier), he discovered that someone else had been vying for Murry's affections. That person was Frieda. Clearly, the three people returning to Kiowa in 1924 (Lawrence, Frieda and Brett) and the fourth who promised to do so (Murry) shared a tangled web of relations reminiscent of other Bohemian circles of the time (such as the one centered around Frieda Kahlo and Diego Rivera in Mexico City). It was not the first time that the Lawrence's Island scheme had been threatened from within by sexual or professional jealousy, but it was the first time that Lawrence and Frieda had isolated themselves with one other woman instead of a couple. One can only imagine that the third party, however enamored she was of Frieda's husband, provided enough companionship and distraction from marital conflict that on balance, her discipleship to Lawrence was something that Frieda decided to tolerate.[7]

Thirty-nine years old, Brett had been a virgin before her love affair with Murry. Perhaps this explains why she thought and hoped he would marry her. Once convinced that there was no such prospect, Brett quickly became a Lawrence disciple. Compared to Katherine

Mansfield, she was much better able to tolerate Lawrence's irascibility both in her regard (he criticized her painting) and in regard to his own marriage (she described a few of the physically and verbally abusive fights witnessed between the Lawrences). Perhaps her tolerance came also from her rather advanced deafness. Although Brett used an ear trumpet, Frieda believed that she used her deafness as an excuse to ignore Frieda's criticisms. At Kiowa, though upper-class Brett certainly didn't lack the money with which to pay rent, she contributed amply in services. She typed Lawrence's manuscripts, hunted rabbits, and helped with the heavier outdoor chores that Frieda avoided. Brett lived in the smallest of three primitive buildings on the ranch, the Lawrences occupied another, and they hoped to fill the third with a good Island candidate. John Middleton Murry was one, and had been involved with both of the earlier Island schemes, but Lawrence was neither surprised nor angry when he backed out and married a twenty-three-year old Englishwoman instead. The initial plan had been to leave the third house open for Mabel and her Indian husband Tony Luhan, but their first visit dissuaded them from returning. As owner of the property, Lawrence forbade Tony to hunt, offending both him and Mabel. Lawrence and Mabel argued privately until Mabel ended up in tears, a rare event that she later characterized as intentional, as her "great gift" to Lawrence (Ellis 187). It was a bit of a performance, too, since Frieda, Brett, and Clarence Thompson all witnessed it.

A month later, Mabel hosted a party at her own house involving drink, boisterous dancing, threats of a fist fight, and Clarence Thomas's warning to Mabel that Lawrence intended to kill her. So *this* was the Island Dream — the perfect setting, the married leader, the steadfast disciple, the imperious doyenne on the outskirts, and the vacant cabin waiting for an inhabitant — a rather poignant picture of how impossible the dream was becoming. The only true disciple that the Island had ever had, the Honorable Dorothy Brett, was evicted from Paradise in January 1925, when Lawrence and Frieda finally decided that she caused more tensions between them than she resolved. Shortly thereafter they returned separately to Europe, where they lived in various abodes in Tuscany and Liguria for the remainder of his life. Brett returned to Kiowa and Taos, continued painting pictures of Indians in "naïve" style, remained close to Frieda, wrote an account of her friendship with Lawrence, and called Taos home for the remainder of her 93 years. Her relationship with Lawrence was the most significant of her life.[8] The fact that Frieda also returned to Taos after Lawrence's death, to live with her third husband, Angelo Ravagli, suggests that some spirit of the ideal community continued beyond the lifetime of its inventor, and was very much in the spirit of the times.

The last mention of Rananim in Lawrence's correspondence appears in a letter to Samuel Koteliansky dated January 4, 1926 and sent from Spotorno, Italy: "The ranch [Kiowa] still doesn't attract me, though sometimes in my sleep I hear the Indians drumming and singing. I still wish my old wish, that I had a little ship to sail this sea, and visit the Isles of Greece, and pass through the Bosphorous. That Rananim of ours, it has sunk out of sight" (v 367). If Rananim dies in 1926, Lawrence's idealism does not. Each of his major late works, *The Plumed Serpent*, *Lady Chatterley's Lover*, and *Etruscan Places*, develops its own version of a little paradise on earth. Although I disagree with Michelucci's characterization of the Mexicans portrayed in *The Plumed Serpent* as a "happy, perfect community," I do agree with her characterization of his changing ideal of Rananim after America (31). Place, indigeneity, and tradition replace personality and *Blutbruderschaft* as primary components of community. Lawrence realizes that "no fundamental change in society can stem from an abstract, superimposed ideal of perfection, seen as universally valid, exportable, and applicable everywhere, but it has to derive from a harmonious, profound contact between the community, its traditions and roots and the surrounding environment, and therefore it varies from place to place" (Michelucci 31). Besides the unstable interpersonal dynamics in the Kiowa Rananim, perhaps the desires of three white Europeans to know and imitate the Pueblo Indians contributed to the instability of their enterprise. Brett, Frieda and Lawrence were not indigenous, had discarded their own cultural traditions, and had no profound contact with the Indian community, which in any case tended to exclude rather than incorporate Whites. Anthropologist Ruth Benedict studied Pueblo Indians at the same time as Lawrence and Mabel. She was very aware of the costs of acquiring esoteric lore for Whites' uses: informants she hired were often ostracized or killed when their "betrayal" was discovered by fellow tribesmen.

Does Island-Dreaming Mean I'm Nuts? Psychoanalytical Approaches to "The Man Who Loved Islands"

Although all of Lawrence's work addresses two competing psychic drives, one for connection, the other for isolation, the story "The Man Who Loved Islands" (1926) does so most consciously and intensely. D. H. Lawrence called it one of his favorite pieces and David Ellis, "the most philosophical fiction he ever wrote" (310). Because he bought three islands, two in the English Channel and one in the Inner Hebrides, actor-writer Compton MacKenzie provided the factual outlines of the plot. But the philosophical questions—what does isolation do to people, and from what does the desire for isolation stem—address Lawrence's own concerns. "The Man Who Loved Islands" is not intended to psychoanalyze his friend—at least the majority of critics do not think so.[9]

An allegory is a literary device in which characters, objects, or events symbolize ideals or moral principles. I will argue that the character Cathcart, and his experiences on three islands, represent three competing ideals within Lawrence—community, marriage, and independence. These fluctuating ideals trouble his shifting attitudes to both the geographical isolation of his Island scheme and to personal isolation as a recurrent problem for him. During a lifetime, many individuals proceed from states of being more social to less so. In the story, and to a degree in Lawrence's life, this pattern emerges. The first Rananim had been sparked by naïve belief in his fellow man and his country; the second Island scheme was founded on his faith in an everlasting friendship with one good man (and his wife); the third utopian community sprang from his desire to share a natural paradise with like-minded individuals. In the last four years of his life, after the publication of "The Man Who Loved Islands," Lawrence felt more alone, and did not try for another Rananim or Island scheme.

"The Man Who Loved Islands" begins in the tone of a fable, which foreshadows a moral to be drawn at the ending: "There was a man who loved islands. He was born on one [England], but it didn't suit him, as there were too many other people on it, besides himself. He wanted an island all of his own: not necessarily to be alone on it, but to make it a world of his own" (WWRA 151). With our awareness of the shaping role that his Island scheme had on his life, we sense right away that these words are a parody of Lawrence's island dream, both for its occasional "androphobia" ("too many people") and its expression of his own megalomania ("to make it a world of his own"). The tone is one of irony — it questions the absurdity and egotism of one man trying to make an entire world. To that irony is added a tone of ominous portent: "[an island] has to be really quite small, before it feels like an island; and this story will show how tiny it has to be, before you can presume to fill it with your own personality" (WWRA 151). The word *personality* as used by Lawrence always signifies his derision, since he believes that personality is only skin-deep — consequently, not the proper focus of either relationships or psychoanalytical study. For him, there is something untrue and superficial about a personality, so we know right away that his islander is deluded about his own true nature.

Fortunately for his dreams, this islander (whose name is Cathcart) had money, so in short order he found himself an island and bought it. To this island, he brought fourteen employees to run the farm and the yacht and the manor (for Cathcart was "to the manor born"). Cathcart divided his time between writing a book on flower references in classical literature and helping his bailiff establish a farm. For about a year, he enjoyed a sense of industry and even contentment, but things gradually began to go wrong. A cow died, a man broke his leg, the pigs sickened, the yacht got smashed, the mason hated the butler, and many of the employees gradually lost their respect for their Master. In addition, the costs of the farm and house far exceeded their profits, and when half of the servants left, Cathcart fired his bailiff and his butler, who had been stealing from him all along. Cathcart began to have bad dreams about the malevolent spirit of the island. In the fifth year, he sold it to a real estate developer for hotels and golf, and moved to a smaller island.

The second island ran on a smaller scale. No longer was he Master; no longer did he "struggle for anything" (162). He brought five of his "faithful" former employees. He continued to write his book, and experienced a different kind of contentment: the absence of desire. He asked himself whether this might be the true nature of happiness. Because Lawrence is famous for his conviction of the absolute necessity for desire, since desire can "resurrect" souls out of their mechanistic death-in-life, the islander's feeling that "no desire spells happiness" indicates he is going in exactly the wrong direction. In his visits to the mainland, Cathcart felt an increasing distaste for the "vulgarity" of city people, since he himself had "dropped out of the race of progress" (163). Inevitably, a serpent entered his new paradise. His housemaid developed a crush on him, and he succumbed to her desire. But love making shattered his pristine aloneness, disgusted him, and "smirched and spoiled" his island (165). He would fulfill his social duties—marrying and providing for her since she was pregnant—but fulfilled his own deepest longings by buying a third, more remote, island, and sailing away, alone.

The third island is a barren rock, which is how Cathcart wants it. On this island, both his personality and his sanity gradually disintegrate. So intent is he on maximizing his isolation that he has the sheep removed, discards the label on his paraffin stove because the letters feel like an intrusion, and is glad when his cat disappears (perhaps it hops the mail-boat to escape from him). The two fishermen who bring mail and supplies disgust him, and he dreads their periodic return. Luckily (in a sense), winter soon comes and kills him before they can return. Hypothermia and psychological dissolution combine to destroy his survival instinct. Though the narrator does not describe his death, he leaves him in such a state of insanity that the conclusion is foregone. Even if he doesn't die physically, spiritually he is gone. The moral of the story shouts at us, without any need to be written: No Man Is An Island.

Critical interpretations of this story have alternated between thinking the sick man Cathcart represents the sickness of modern Europe, or that he is a human aberration, representative only of illness. In other words, these critics are debating the question of how closely does the narrator of the story identify with his subject, Cathcart? This question can also be asked about T. S. Eliot's famous poem of a radically isolated man, "The Love Song of J. Alfred Prufrock" (1917). In the poem, is it Prufrock's timidity that effectuates his demise, or is it the hypocrisy of the social forms, figured in the ladies' tea party? The best answer to both questions is "a little bit of each." Because of its cost to psychic wholeness and its wasteful byproducts in the environment, the race of progress takes a toll, even as it produces great artifacts of culture and industry. Paradoxically, Cathcart withdraws to islands because of his environmental conditioning, his involvement in the race: he believes he has found an environment over which he can have total control. In his debate with Bertrand Russell over what "subjectivism" meant, Lawrence argued that community must be based on a shared religious vision, while Russell thought that, in matters of the soul, it was every man for himself. Cathcart demonstrates Lawrence's fear of "every man for himself" taken to the nth degree. Cathcart's will to power is already extreme on the First Island, and only intensifies on the Second and Third. So, the islander is both an Everyman demonstrating the *reductio ad absurdum* of the selfishness caused by industrial capitalism, and a case study in abnormal psychology.

Psychoanalytic critics of the story have taken various approaches, but each of them diagnoses a state of psychotic regression in the islander. Ben Stoltzfus performs a sophisticated Lacanian analysis of the islander as an individual regressing, or attempting to regress backwards through the Mirror Stage.[10] Cathcart desires to dwell, not in the Order of the Symbolic, where adults must spend most of their time, but in the Order of the Imaginary, where the subject does not recognize separation between itself and the world, itself and the mother. Stoltzfus says that each of the three islands can be taken to represent each of Lacan's three Orders, or parts of the psyche: the Imaginary, the Symbolic and the Real. The Mirror Stage is a fundamental passageway between the Order of the Imaginary and the Order of the Symbolic. It occurs between the ages of six and eighteen months, when the child sees its own reflection in the mirror and begins to conceive of itself as a unified being, separate from the rest of the world. At the Mirror Stage, the child gradually passes from the Imaginary to the Symbolic Order, primarily through the acquisition of language. As language is a system concerned with lack and separation (since it names what is not present and substitutes a linguistic sign for it), maturation does not occur without a sense of breakage and loss. The third Order, that of the Real, is the most difficult to describe, as it is the realm of unconscious desire. It is literally the physical world and, symbolically, all that a person is not.

According to Stoltzfus, what happens on the first island represents Cathcart's engagement with and rejection of the Symbolic, which is the realm of law and social opinion. When his island employees form a bad opinion of him, he does them one better and rejects them by selling the island. On the second island, he explores the realm of the Imaginary. Although the child does not wholly realize its difference from the mother in this stage, it still feels her absence, so it must "come to terms with the postponement of satisfaction, the repression of desire, and the nurturing of discontent" (Stoltzfus 28). On this island, Cathcart postpones satisfaction with both his work (book) and his sexual relationship, while nurturing the discontent ("other people are so dirty") that drives him finally to the third, stark island. The third island represents his encounter with the Real, consisting of his unconscious relationship with death, and with it, his desire to return to the mother's womb. On each island, Cathcart's ploys to achieve isolation are attempts to revisit the psychic site of his primal wounding, the self-splitting implied in the Mirror Stage. He discards language (the label on his paraffin stove and all other printed material) and other forms of connection to the Other, in hopes of reenacting the prelinguistic bond with the mother. This presents obvious problems—an adult psyche cannot survive such a psychic "experiment"—but what is worse, it leads inevitably to death. The end destination of psychic regression, the return to the womb, is death. As Stoltzfus sums up this story, "to love 'I-lands' is to dwell within the split self, a division that mimes the splitting during Lacan's 'mirror stage' of the infant's development" (27). This analysis helps to explain Cathcart's anger and horror; he is reliving his primal cleavage with mother and all the goodness that she signifies.

John Turner performs an equally sophisticated Winnicottian analysis of Mr. Cathcart.[11] The islander is a man who "needed solitude but who lacked the capacity to be alone," he begins (259). Psychoanalyst Donald Winnicott theorizes that the infant develops the capacity to be alone by retaining a sense of mother's presence even when she is absent. The growing child gradually acquires the ability to de-cathect the mother's presence and spread its essence — the benign — throughout the human and natural world. This "belief in a benign environment" strengthens the ego so that it can withstand shocks from the other side of the unconscious (what Jung calls the Shadow). Relationships and experiences that inspire terror or hatred will strengthen a strong ego, because the psyche learns to integrate and accept its own ambivalent feelings. Cathcart experiences these fearful and hostile feelings on the first and second islands, but his weak ego, or schizoid personality, cannot integrate its negative feelings and impulses. One common defense to such a problem is isolation.[12] By avoiding others, the person tries to run away from his problems. The schizoid psyche engages in "magical thinking that is addicted to the novelty of fresh starts," just as Cathcart is caught in a repetition compulsion by continuing to buy islands even when they are proving disastrous (266). He denies his root problem, the failure to acknowledge and integrate his own negativity. What typically causes such a weak ego? A mother who was either too possessive or too neglectful to "create in her child the belief in a benign world where he may safely find himself" (266). We must be careful not to mistake Cathcart for Lawrence, but, after a few months of marriage, Lawrence began to blame his own over-dependence on his wife to the over-protectiveness and possessiveness of his mother.

Critic David Willbern extends this discussion of Cathcart's fundamental negativity, adopting Freudian concepts of orality and ambivalence in his analysis of Cathcart's diseased psyche. In "Malice in Paradise: Isolation and Projection in 'The Man Who Loved Islands'" he calls attention to the seething rage that the islander experiences on his islands. Cathcart even projects his own lust onto the island. He dreams that it contains a malevolent spirit. Noting the prevalence of images of nurturing orality on the first island ("two humps of hills," "cows lay chewing," "cosy and home-like"), Willbern finds they shift to a destructive, devouring orality as soon as things begin to go wrong. The island "swallows" his money, the sea "sucks angrily" and "eats at" the land. Willbern compares Cathcart's ambivalence towards Flora to Lawrence's ambivalence to his cow Susan, whom he tended at Kiowa. Lawrence loved the contentment he felt while milking her, but feared something "demon-like" in her when she ran away and hid among the trees. Like an infant in the oral stage of development, he felt rage and distrust when he could not keep that milk-bearing presence nearby. Willbern thinks that Lawrence's and Cathcart's ambivalence about orality — the fact that what nurtures can also destroy — "lies at the base of ... [their] radical ambivalence about intimacy" (229).

The prominent component of the psyche stuck at the oral stage is sheer rage at its failure to control its world, which in the case of the infant means access to the nurturing qualities of the mother. Cathcart's psyche abounds in such feelings, yet he projects them onto the outside world instead of owning them as his own. His first island sucks his money like "an octopus with invisible arms stealing from you in every direction" (WWRA 158). Life might seduce you with hope, but you are always doomed: "As sure as the spirits rose in the human breast, with a movement of joy, an invisible hand struck malevolently out of the silence" (158). Similarly his paranoia about the "molestations" of the fishermen on the third island is a projection of his own hostility (Willbern 232). This hostility arises when the wish for nurturance arouses memories of lust, disappointment, and loss. The projection of those feelings onto the environment creates such a hostile outer world that the psyche fears and retreats from it. However, isolating oneself only brings the psyche into repeated confrontation with the fears and hatred that are its own, what Freud calls the "return of the repressed." Paradoxically, isolation exacerbates the very feelings it is intended to avoid.

Freud, Winnicott, and Lacan — as applied by literary critics Stoltzfus, Turner, and Willbern — cast an illuminating light on island quests. Is everyone with an island dream at base an angry man or woman guilty of the magical thinking that a fresh start will heal his or her primal wound of separation from the mother? These studies highlight the fascinating psychic economies at the heart of psychoanalysis. In order to defend itself, the psyche often constructs elaborate fantasies and actions "necessary" for self-preservation. Instead, these protective devices ultimately destroy it. Cathcart is the perfect example. In order to protect himself from the hostility of the world, he isolates himself on islands. Without any distraction or help from the outside world, his diseased psyche gradually destroys itself.

Lawrence was more aware than most that his upbringing had resulted in an unresolved Oedipal complex. One of Frieda's greatest attractions was her knowledge of Freudian thought and her attempts to lift her own repressions. Lawrence's contacts with his friends and wife often caused him (and them) a great deal of pain. However, I believe it was a sign of his emotional health that he persisted in seeking new relationships and continuing old ones, even after Rananim as such was dead. To read his biography is to encounter a phenomenally large cast of acquaintances, friends, and disciples. Lawrence was once accused of not letting go of relationships that had already expired. The fact that he tried to pick up several relationships where they had left off, after giving both offended parties time to cool down (such as Bertrand Russell, John Middleton Murry, and ultimately, Frieda) suggests that, unlike Cathcart, he did not burn his bridges. All of his novels stress the necessity of connection, even if his protagonists have first to isolate themselves in order to understand it.

Utopia, Sex, and the Lone Wolf

Like many allegories and fairy tales, "The Man Who Loved Islands" comes in three parts, each with a separate lesson. Lawrence subtitles the sections "First Island," "Second Island," and "The Third Island." The neatly tripartite structure recalls the powerful symbolic values associated with the number three. The Holy Trinity greatly interested Lawrence as a symbol of alternatives, of breaking the binary opposition of Son and Father, of senses and mind, of self and "not-self," in short, of two ways of knowing. In *Twilight in Italy* Lawrence explains that the Holy Ghost "relates the dual Infinites into One Whole" (46). There is something whole or complete in the number three. It is a powerful number in folk wisdom: "trouble comes in threes," "three times a charmer," and "give someone three chances." Three is a common number in both epics and fairy tales. Dante encounters three beasts; Goldilocks, three bears. In rhetoric, three examples serve to prove most points. Cathcart's three islands are clearly significant allegorically, not only by virtue of their number, but by their decreasing substantiality as "worlds" he could "own"—community, the sexual Other, himself. Like a fairy tale or parable, this story conveniently isolates the main character's three lessons, each one a little more vital to his survival. Unlike in a parable however, this story's hero never learns. The story's ending resembles the most fatalistic ones of Grimm's Fairy Tales or Aesop's Fables, where we learn from the misfortunes— including death — of others.

The three islands stand for three ideals that fluctuated in importance in Lawrence's mental life: the Island scheme, sexual resurrection, and independence. These three were not mutually exclusive; he mixed and matched these ideals in varying proportions, like a wardrobe for right living. Yet there was a general progression from the first to the last over the course of his lifetime. Unlike utopian fiction, "The Man Who Loved Islands" does not present the shiny side of these three ideals, but rather their Shadow side, their propensity towards destruction. A community based on autocracy will implode. A sexual union lacking respect and communication will fester. A person who avoids all others will die. Showing the down side of his ideals signifies that Lawrence is engaged in self-analysis and self-parody (there is something funny about the extremities of Cathcart). The negative capacities of Utopia, sex, and isolation drive the work, enabling us to appreciate the unstable nature of all ideals.

All allegories have a moral, and I have claimed that this allegory teaches that no man is an island. This lesson becomes most apparent on the third island, but what do the other two islands teach? The fundamental problem with his relations on the other two islands is that they are not equal. Lawrence believed in "natural aristocracy," the idea that people are not created equal, and that the ones with naturally finer sensibilities should lead the rest. However, that very sense of *noblesse oblige* in Cathcart caused the people to turn against him. Although they respected his status as their Master, theirs was only an outward show of deference, nothing profound. The absence of real give-and-take relations between people sharing a home leads to its demise:

> It is doubtful whether any of them really like him, man to man, or even woman to man. But then it is doubtful if he really liked any of them.... He wanted them to be happy, and the little world to be perfect. But any one who wants the world to be perfect, must be careful not to have real likes and dislikes. A general good-will is all you can afford.
>
> The sad fact is, alas, that general good-will is always felt as something of an insult, by the mere object of it; and so it breeds a quite special brand of malice. Surely general good-will is a form of egoism, that it should have such a result! [WWRA 156].

Although he believed in natural aristocracy, these paragraphs show that Lawrence knew that most people don't. The employees placate Cathcart's need for respect, yet their economic need compels their stance of deference, thereby killing any natural respect.

The ideal that inspired Cathcart's farm on the first island is far different from those that led to Lawrence's Raninim. Cathcart merely wants to "make a world" for himself, out of his need for dominance and his dislike of "too many people." Lawrence wants to make Rananim for others, too, so that they may *rejoice* in God's world. His and Bertrand Russell's Rananim was calculated to fix the problems of England, not to retreat from them. The early Rananim seems altruistic, though one might argue that wanting to be leaders of their utopia implied an immense will to power. The striking similarity between the first Rananim and the First Island is that they were to be perfect worlds. As the satirical tone of the passage about Cathcart's generalized good will suggests, by 1926 Lawrence regarded the urge to perfection inherent in his first Rananim as a form of megalomania. He was coming through the other side of his Leadership Period, realizing that relationships based on power are not "real," in the sense that they presume the superiority of one man over others on the basis of class or other indices of status. Lawrence's natural aristocracy was not based on social class, but on a personal classification of integrity of soul and body. This story inverts the paradigm, and in doing so, teaches a lesson. Cathcart may think he is a natural aristocrat, but his failure to connect proves otherwise. In fact, Cathcart is the very antithesis of Lawrence's natural man since he is so out of touch with himself, others, and the cosmos. So, any obedience he achieves is necessarily only economically driven, and therefore false, as it does not come from any real sense of connection.

Can this obstacle to success on Cathcart's first island be related to Lawrence's first Rananim? Were the relationships between the participants based on economic or social hierarchies rather than real connections? Quite possibly, the class difference between Lawrence on the one hand and Lord Bertrand Russell and Lady Ottoline Morrell, on the other, caused the disruptions that proved fatal to the project. Although it is certainly not a part-to-part analogy to Cathcart and his employees, their radically different way of seeing and talking about the world prevented Russell and Lawrence from collaborating on their lectures. In his style of thinking and speaking, Lawrence was highly influenced by his Low Church background. Even after he had rejected Christianity as a belief system, its value as a mythology and rhetorical style continued to influence him for his whole life. He uses both the images and the style of the Bible throughout his entire life. Russell, on the other hand, came from a distinguished line of aristocratic social reformists. His grandfather Lord John Russell had helped to draft the Reform Act of 1832 and resigned as Prime Minister on the defeat of the second parliamentary reform bill (KW 188). Social reform through parliamentary channels was in the bloodline. Reason, not religion, ruled this family, which had the social status to insert itself into mainstream politics and institutions.

More likely, Cathcart's little kingdom reflected a more extensive period of Lawrence's life than the first, failed Rananim. The middle years of Lawrence's career, from about 1921 to 1926, came to be known as his Leadership Period, since his writing of the time espoused the belief in rule by natural aristocrats. *Kangaroo* dissects the rise and fall of a fascist party in Australia; *Aaron's Rod* explores one man's submission to a natural leader on an interpersonal level; and *The Plumed Serpent* resurrects pagan culture in Mexico through a political movement headed by a natural aristocrat. Lawrence's espousal of these values was by no means absolute. In these novels, female characters struggle against the hegemonic injustices of this philosophy. Even the leaders themselves are easily deconstructed through their own internal monologues. Therefore, the novels and stories can be understood as dialectical discourse; they engage the struggle between opposite poles of egalitarianism and authoritarianism. After 1926, Lawrence redirects his attention to the domestic and historic spheres, lessening but not relinquishing his desire for political and personal hierarchies. Cathcart's failure on the farm is due to his erroneous belief that he could own a world, and reflects Lawrence's self-criticism about his own politics of hierarchy.

The Second Island represents Lawrence's second ideal of sexual union, but on its dystopic side. Lawrence is one of the world's greatest advocates for marriage: "Man or woman, each is a flow, a flowing life. And without one another, we can't flow," he writes, in "We Need One Another" (*Phoenix* 192). This flowing together is best conceived as a life-long relation, regardless of the fact that "sex-desire departs," at times. Lawrence calls the whole relationship of marriage by the word "sex," indicating that the long-term interplay between the two genders matters much more than the sex act: "[T]he great flow of the relationship goes on all the same, undying, and this is the flow of living sex, the relation between man and woman, that lasts a lifetime, and of which sex-desire is only one vivid, most vivid, manifestation" (*Phoenix* 194–5).

Besides marriage, sex needs tenderness. *Lady Chatterley's Lover* proclaims the crying need for sensitive touch. As Cathcart knows, sex without it is horribly degrading. Why can he not feel anything with Flora? Like all relations on the First Island, the relationship to Flora fails because it is not real, not being based on equality or respect. Flora's unquestioning submission to Cathcart, and her ability to absorb his unspoken hatred, are the very things that he most loathes about her. He describes their sex as mechanical and automatic. "Tenderness," that shibboleth of *Lady Chatterley's Lover* which makes Oliver and Connie's love different from and better than others, is not in Cathcart's physical or verbal vocabulary. Part of the problem is projection: not able to believe that anyone could love him (for he does not love himself), he thinks that Flora's love is a mere act of will: "[I]t seemed to him, as far as her physical self went, she had not wanted it either. It was just her will" (MWLI 164). Particularly distressing is her mute submission: "She served him with the same silence, the same hidden hunger to serve him, to be near where he was. He felt her love following him with strange, awful persistency" (164).

Lawrence's marriage was not like this, nor did he want it to be. He believed in the free expression of feelings, however much tension they produced.

Perhaps Flora is the ultimate self-parody. Lawrence once wrote that a woman's whole purpose in life is to serve her man. In *Fantasia of the Unconscious* (1921), Lawrence's treatise on depth psychology, he explains that while all men need both marriage and work to fulfill themselves, women do not need this secondary goal. The whole purpose of a woman's life is to support her man, believing in his "greater purpose" (FU192). What would such a woman look like? Like Flora — a mute, yet needy pest to a man who does not have a greater purpose, and does not love her. It is not a question of Flora being the wrong choice for Cathcart, nor a question of her "personality." For Cathcart does not wish to find a better relationship, only to have no relationship at all. For Lawrence, marriage was the single most important relationship of life, without which man and woman cannot even claim to have lived. Rejecting sexuality puts Cathcart on the slippery slope to death.

If the Second Island is a self-parody of Lawrence's sexism in believing women are the second sex, does this imply that he later discards this belief? *Lady Chatterley's Lover* (1928) offers some insight. Connie Chatterley is a natural aristocrat as well as a social one. She has a profound sensitivity to what is wrong and right in her world. Her lover, on the other hand, harbors a profound distrust of women, exacerbated by a marriage in which his wife knew how to have an orgasm without him. This made her a "Lesbian ... who grinds her own coffee" (LCL 219). Mellors combines aggression and tenderness in his love-making; his anger at strong women in his past sometimes clouds his trust of Connie. But Lawrence balances Mellors's anger with Connie's response to him, showing he is aware that this character is extreme and has the capacity to hurt others. She calls him on his sexism and misogyny, as in this passage where he is raging against his soon-to-be-ex wife:

> MELLORS: When a woman gets absolutely possessed by her own will ... then it's fearful, and she should be shot at last.
>
> CONNIE: And shouldn't men be shot at last, if they get possessed by their own will? (LCL 280).

In all of Lawrence's fictions, the neuroses of the main characters are identified, contradicted, and at least partially healed by the others. Lawrence uses dialectical discourse between characters to modulate the extremity of his ideas. He consistently symbolizes the opposites within himself in characters who represent his different aspects. The fact that he writes extremely convincing females is proof of his androgynous imagination, which Virginia Woolf identified, in "A Room of One's Own," as the sign of a true artist.

The Third Island represents Lawrence's ideal of individualism, of standing alone in the face of a sick society and a domineering wife, but it shows nothing of the positive value of the ideal. Certainly Lawrence had written amply about the importance of individualism, particularly about knowing oneself in the face of social movements that threatened to destroy independent thinking. In newly liberated post-war Italy, Lawrence was disgusted by the "unmanly insolence of these lords of toil," as workers claimed their rights; "save us from proletarian homogeneity and khaki all-alikeness," he prayed (SS 89). Think for yourself, could be his motto. But Cathcart is not a thinker, and his isolating moves are not intended to prove any political or social points. On the contrary, as the psychoanalytic critics show, his moves are driven by the unconscious, and remain unanalyzed by Cathcart's conscious mind. Therefore this ideal of individualism is shown only on its negative side. Is Lawrence saying that too much individualism inevitably leads to isolation, as he frequently feared it did for him? For dangers surround the isolated individual like a sea surrounds an island. This Third Island represents the death-like tug of fear and misanthropy, out of which Cathcart's desire for isolation stems.

Lawrence's own life contains a far stronger tendency to the opposite impulse, that of human connection. He was very aware of his own "lone wolf" tendencies, and of both the ideals and the psychological complex (Oedipal) that kept him separate. The biographer of his middle years believes that his stance somewhere between anti-war and pacifism, as well as his courageous proclamation that every man is inherently bisexual, isolated him from many potential friendships during the war years (KW 192–97). The biographer of his late years, David Ellis, performs an excellent analysis, not psychoanalytical but psychosocial, of the extent to which Lawrence was or was not "properly socialized." It is worth quoting at length because it provides an expert's answers— Ellis having spent the most time with the materials— to the question that continually arises in Lawrence studies: what kind of friend was he?

> Although he learned some caution in later life, there was never far below the surface in Lawrence the impetuous urge to go out to others in search of a genuinely warm human contact. As much as any awareness of his exceptional talents, this quality, which rightly or wrongly Lawrence himself associated with his working-class background, explains why so many people who met him for a short time recalled being in his company with such pleasure. He never learned that art of manoeuvring among others with an eye constantly alert to one's own psychological advantage.... To a large extent he remained child-like, although not childish, in his search for love. Yet those who agree with him in thinking real human contact ... [is] one of the few substantial happinesses of life will not be inclined to call him immature, or dispute his claim that it was the general state of life in society which deserved to be considered 'abnormal' rather than his own. One could say that on many occasions he egotistically expected too much of others, and too much of life ... but one would have to be curiously eupeptic not to be affected by his analysis of the unfortunate way we live now... [370].

It is noteworthy that Ellis characterizes Lawrence as constantly reaching out to others, rather than pushing them away. Lawrence was not an excluder or a snob; he never refused anyone entry into Rananim, almost never ended relationships with finality, and had no class snobbery, either upwards or downwards. His frequent feelings of isolation were not caused by outright rejections of others, but by the effects that his individualism had on them, and on occasion, by the vehemence with which he criticized their ideas. Yet Lawrence never intended to isolate himself. On the contrary, community and marriage were the two constants in which he believed throughout all the other vacillations in his philosophy of life.

To the extent that Cathcart "represents" Lawrence, then, he probably could also be said to represent every soul in crisis. The Lacanian, Winnicottian and Freudian analyses of the story reveal the islander's psyche to be suffering from its unresolved traumas of infancy and childhood. Ellis describes Lawrence as child-*like*, not child*ish*, in his expectation of love. The adjective suggests someone who is able to absorb past rejections and tensions, and persevere in strengthening other relationships while adding new ones. To be childish, on the other hand, would be to demonstrate more of that infantile rage at the absent mother, that interpretation of the universe as malignant, and that self-enclosure which we see in Cathcart.

Lawrence was intrigued by the phenomenon of the "persona" and the way in which it could contribute to psychic isolation (a tragedy which Matthew Arnold described in his poem, "The Buried Life"). In 1927, he reviewed American psychoanalyst Trigant Burrow's new book *The Social Basis of Consciousness*, which argues that "nearly all human behavior is determined by the effort of human beings to adapt themselves to the images of themselves which their upbringing has persuaded them to accept" (Ellis 369). Because of his scorn of the persona or personality as a basis from which to act or to understand human nature, Lawrence readily accepted Burrow's emphasis, as well as his diagnosis that "so-called normal activity is neurotic" since it stems from the superficial personality. On July 13 he wrote to Burrow diagnosing the cause of his own sadness as social isolation:

> What ails me is the absolute frustration of my primeval societal instinct.... I think societal instinct much deeper than sex instinct—and societal repression much more devastating. There is no repression of the sexual individual comparable to the repression of the societal man in me, by the individual ego, my own and everybody elses [sic]. I am weary even of my own individuality, and simply nauseated by other people's. I should very much like to meet somebody who has been through your laboratory, and come societally unrepressed. Is there anybody? [vi 99].

Characteristic here is Lawrence's willingness to blame himself equally with others for acting out of ego (rather than what—tenderness?). His relative psychoanalytical sophistication at the age of forty-two prevents him from thinking that a "laboratory" experiment can really free one from the paradigm of social repression. However, just as interpreters of the story "The Man Who Loved Islands" alternate between diagnosing the individual and diagnosing the society as the sick entity, so too did Lawrence alternate between faulting himself and society for his aloneness (although more often he chose the latter). In another letter to Burrow dated August 3, 1927, Lawrence blames his isolation on both the unsociable nature of the human being and on his own particular historical moment:

> Myself, I suffer badly from being so cut-off. But what is one to do? One can't link up with the social unconscious. At times, one is *forced* to be essentially a hermit. I don't want to be. But anything else is either a personal tussle, or a money tussle....
>
> There will never be a true societal flow — all things are relative. Men were never, in the past, fully societal — and they never will be in the future. But more so, more than now. Now is the time between Good Friday and Easter. We're absolutely in the tomb [vi 113].

Historically speaking, he was uncannily right, if Good Friday stands for World War I and the devastation it wrought upon nations and individuals. As he wrote these words, Europe festered under new clouds, the various forms of fascism that would not be eradicated from Europe until late in the century. Three-quarters of a century later, we have not risen from the tomb, and Easter is a long way off. Perhaps today we are a little more "societal," as Lawrence says, in that a democratic ethos apparently rules in the United States and England (democracy is the *persona* of these nations, while their actual *self* is somewhat different). However, social ideologies such as multiculturalism and political correctness, as well as programs for international human rights, affirmative action, and gender equality do not necessarily increase what he called the "societal flow." Instead of social or political platforms, he is looking for "real human relations," and finds instead either too little involvement with others (as in "ordinary acquaintance") or too much ("tussles") (vi 113). Nonetheless, he continues to seek connections, quite unlike Cathcart, who has forsworn all human relations.

Putting together the lessons of the three islands, the message that emerges is the profound need for equal and integrative relations with both others and the self. The First Island, by its demise into a scene of distrust and paranoia, teaches the unfeasibility of the wish for absolute control, either control over others or control over geographical space. The Second Island teaches the need for sexual relations to be both loving and equal. The Third Island teaches the absolute need to gain understanding about oneself, including one's neuroses. This lesson is the most fundamental of the three, because forgiveness of the self prefigures the capacity to love others, without which we will surely die.

Lawrence's fictional story, "The Man Who Loved Islands," and Lawrence's life story — his attempts to make ideal community over a lifetime — suggest how common and how profound mental islands are, as representations of competing psychic drives. The paradisiacal island of the mind is a place in which one's societal impulse is fulfilled by perfect relations, and the place where the ultimate confrontation with the self occurs. As Cathcart demonstrates, the psyche both wants this confrontation (to heal the primal wound), and avoids it by various mechanisms such as projection and denial. Paradoxically, isolation is a psychic defense mechanism, yet it also precipitates death. The paradox of islands as both alluring and repelling space creates excellent subject matter for literature. As Melville noted in *Moby Dick*, the soul is aware that its "insular Tahiti" of ideals is threatened on all sides by the "appalling ocean" of the Real.

VI

"The Cure for a Family Is a Family": Margaret Mead and Samoa

> Margaret Mead, Margaret Mead
> Helps to fill our country's need,
> Thinks our culture is much lower
> Than the one that's in Samoa....
> — *Margaret Mead: A Life*

Connections: Rananim vs. the World Family

Like painting and literature, anthropology was a site of contest during the first three decades of the twentieth century. While Gauguin expanded on the credo of Post-Impressionism, where form and color were emotional expressions, Synge opened up the subject matter and language of drama. Lawrence proclaimed a new kind of novel, which emphasized the importance of the unconscious in our psychic and physical lives, and Margaret Mead demonstrated that all cultures are valuable in their own right. In doing so, she challenged the biological and racist bent of anthropology. Although one must guard against facile categories (as the Margaret Mead/Derek Freeman controversy amply proves), a broad characterization of the problem as one of Nature versus Nurture is nevertheless accurate. On the Nature side were the Social Darwinists, whose extrapolation from natural selection theory resulted in such unhappy postulates as the superiority of the white race and the plan for social engineering (eugenics) to make a better world. On the Nurture side were the cultural determinists who would argue that only culture, never biology, determines the behavior of an individual. From the perspective of our own times it is clear that only a middle ground would be reasonable. Before such synthesis could take place, however, the thesis and antithesis needed their time in the light. Margaret Mead took a stance somewhat left of center, if the left pole represents cultural determinism. As she aged, and as the field of anthropology matured, she incorporated the possibility that evolution does play a role in culture formation, gradually substituting a "both/and" for an "either/or" framework in the battle between biology and culture.

Cultural determinist she was not. Her banner — since such was needed in those heady days of early anthropology — was cultural relativism, the proposition that each culture has its own value, irrespective of other cultures. Part of this belief system was methodological; it instructed anthropologists how to approach their fieldwork with sensitivity. Because we always approach other cultures with some degree of ethnocentrism, we must endeavor consistently to open our minds to difference, to suspend judgment, and to learn from the "good" of other peoples, always reflecting on our own cultural values and assumptions. Today in all disciplines of the social sciences and humanities, the practitioner or scholar is expected to consider his or her own subjectivity, asking, "how do my age, gender, nationality, religion, and class affect my interpretations?" As individuals, scholars have varying degrees of consciousness of the subjectivity of their own narratives. On an institutional level, however, the awareness is acute; affirmative action makes this a legal imperative. We see the beginnings of such sensitivity to multiculturalism and diversity in the work of Mead and other cultural relativists early in the century.

Darwin's importance to Margaret Mead's work has already been mentioned. Eugenicists (Francis Galton and Charles Davenport most prominently) appropriated his theory to their own uses, while Mead and the "Boasians" (disciples of Franz Boas, Chair of Anthropology at Columbia University) refuted the eugenicists' unproven theories.

Mead's story also connects to our other islanders in instructive ways. Ideologically, Mead links up to Gauguin in her early career, when she idealizes the Samoan way of life. Like him, she saw what she considered a healthy attitude to sex as the most paradisiacal attribute of "the Samoan way." Like him, she had a tendency to over-generalize, as in this sentence about Samoan sexuality: "Familiarity with sex, and the recognition of a need of a technique to deal with sex as an art, have produced a scheme of personal relations in which there are no neurotic pictures, no frigidity, no impotence, except as the temporary result of severe illness, and the capacity for intercourse only once in a night is counted as senility" (CAS 151).

Very different kinds of connections may be made between Mead and Lawrence. Ideologically and temperamentally opposed, he is Modernism's alienated artist par excellence, while she is one of Social Science's most zealous reformers. Never alienated, Mead is willing to lead the world in social reform. They are alike, however, in their mutual fear of isolation. His desire for community and connection leads him into numerous friendships with other intellectuals; hers leads to extraordinarily extensive networks among intellectuals and reformers, with herself playing the focal role in each system. Whereas Lawrence conceived Rananim and the other Island schemes as an escape from society, Mead created extended families of individuals who wished to recreate the world in their reformist image. Despite their intense drive to belong to something greater than themselves, both Lawrence and Mead appear essentially lonely, with a marked increase in their isolation as they age. Unlike Lawrence's, Mead's fear took on manic proportions; she would not eat alone, live alone, or go alone to the field after the first trip in Samoa. "Be lazy, go crazy" was her childhood motto; its apparent mocking acceptance of the Protestant work ethic masks a real fear of being alone with her feelings. Workaholism and a rather desperate reaching for human contact — at the same time that she also pushed it away — helped Mead avoid her feelings of fear and insecurity, without providing the intimacy she longed for. She worked on such a large scale (as Jane Howard put it, "the world was her field") that it became difficult to cultivate the small, the local, and the intimate areas of her life. When she consoled her friends who were having marital or parental problems with the witticism "the cure for a family is a family," she meant no less than the world as a possibility (Howard 357). The possibilities of relationship and network were endless, and it was this incurable optimism about life's abundance that made her an icon of The World Family. As her biographer remarked, "she made the impossible seem accessible" (Howard xv).

This chapter will examine this curious characteristic of Margaret Mead — the double bind of her need for belonging and her protective, isolating moves—and how her first book prefigures her alternative lifestyle. Her value judgments about the *fa'a Samoa* (Samoan way of life) and its contrast to American society point the way to her life-long efforts for reform. Although she later recognized errors both in her methodology and her conclusions in *Coming of Age in Samoa*, the emotional viewpoint of that book nonetheless expresses her own deepest needs and drives. Esteeming certain aspects of the *fa'a Samoa*, such as the extended families, easy adoptions, and guileless, sexual play, Mead tried to recreate this ethos in her own life. She achieved a much-modified, American version, but not without paying a heavy price. Her essential loneliness suggests the perils of imitating other cultures and affirms one principle of cultural relativism — that an individual's behavior and values are an amalgam of the culture in which s/he lives. I do not mean to say that one is absolutely determined by one's culture any more than by one's biology (race, gender, temperament of parents, etc.), but one adapts to one's own living conditions, not to those of remembered populations halfway around the world.

Mead's objective was to create for herself and others an extended family in America and the world. Her range of interests and the sphere of her influence were so large that, in the 1950s, *Time* magazine dubbed her "the mother of the world." Unfortunately for her, the American ethos puts the nuclear family at its very core. In *Coming of Age in Samoa*, Mead describes the American family as an isolated unit waging ideological war on other isolated units: "But in our society, home discipline is used to establish one set of standards as over against other sets of standards, each family group is fighting some kind of battle, bearing the onus of those who follow a middle course, stoutly defending a cause already lost in the community at large, or valiantly attempting to plant a new standard far in advance of their neighbours" (241). The problem is not only the sense of competition with one's neighbors, but also the isolation that results. In urban and suburban America, Mead diagnoses a tragic failure of community. In one of her many amusing epigrams, she notes, "Nuclear families, isolating everybody as they do, are scarcely less dangerous than nuclear bombs" (Howard 355). The statement displays many constants about her character: her exaggeration, her black humor, her gutsy frankness, and her only slightly masked, abject fear of being alone, trapped in an institution not of her making.

In light of her desire to make an extended family out of her peers and the world, this chapter will examine Mead's early Samoan experience, the conclusions she drew from it, and her attempts to apply those conclusions to her own life. The isolation of Ta'u, her island and village in the American Samoa archipelago, was both good and bad for fieldwork. The island's isolation was good for anthropology because it preserved intact a culture that, though Christianized by missionaries, retained its primitive (non-industrial, non-capitalistic) way of life. Mead's isolation on the island was bad for her career because it lead directly to errors that caused the Mead/Freeman controversy. How was she to collect information about sex from non–English speakers, and synthesize it into conclusions that proved or disproved her (and Boas's) working hypothesis? Academic work can be isolating, and anthropology in Mead's time was particularly so, when scholars were rushing to claim a "people" for their own before primitive lifestyles disappeared (historian George Stocking called this the "my people" syndrome). Add to this the fact that women were patronized in this male-dominated field, and assigned to work not only in "safe" environments, but also on "female" topics. Mead felt that Anthropology adopted a tough-guy approach to fieldwork (one that I believe occurs in most other professions as well)—the beginner is expected to cut her teeth on her first large project without extensive mentoring. Perhaps this is a vestige of the initiation rite of primitive societies.

Upon her return to New York from Samoa, Mead's behavior patterns reflect her ideal of extended family in both work and home settings. She collaborated with social scientists from many fields, psychology being a favorite choice. She tried to keep all of her homes from being nuclear, because of the volatility and loneliness she associated with that way of life. The second half of her life, from 1948 on, gives a picture of increasing fame and decreasing intimacy. In her comparison of American and Samoan sexual mores, Mead deems American methods of gaining a "more dignified standard of personal relations" (than primitives appear to have) as "exceedingly expensive," saying that intense personal attachments can cause pain and neurosis (*CAS* 211). Perhaps the manic busyness of her later years proves that the price one pays for having an alternate social vision is also precariously high. Mead's story provides a fitting closure to this study of modernist innovators on islands because, in accentuating both positive and negative uses of islands, it picks up recurring themes. Positive island experiences, such as Mead's affirmation of the *fa'a Samo*a, are often idealized into life-guiding principles. Negative mental islands — those abject fears of personal isolation — exert strong influences on both conscious and unconscious behavior, often leading to the very condition one sought to avoid. This double pattern is the enigma of Margaret Mead, and the dilemma of all artistic and intellectual iconoclasts.

Mead Comes of Age in Samoa and Columbia University

Coming of Age in Samoa is the top seller in all of anthropology. It is a must on freshman anthropology syllabi and appeals to the general public as well. At the age of twenty-seven, Mead performed the unexpected miracle of popularizing anthropology, with her very first publication. It is no wonder that she is perceived as The Giant in her field, and that certain male anthropologists would like, symbolically and professionally, to kill the Giant. While Lowell Holmes admits to this desire in the opening pages of *Quest for the Real Samoa*, he nevertheless proceeds to treat Mead and her Samoan work with a great deal of professional respect. Another colleague in Samoan studies is not such a gentleman. Derek Freeman, erstwhile Professor of Anthropology at Australian National University (he died in 2001), made a career out of giant busting. His two books, *Margaret Mead and Samoa: The Making and Unmaking of an Anthropological Myth* (1983) and *The Fateful Hoaxing of Margaret Mead* (1999), indicate clearly by their titles that his mission is to discredit Mead. If Mead in later years was manic in her busyness and brusque with her colleagues, Freeman became obsessive in his attempts to discredit her foundational work and hostile in his responses to her defenders. The ongoing dialogue between defenders and detractors of Margaret Mead has been called science's Liveliest Debate of the Century.

What accounted for the popularity of *Coming of Age in Samoa* when it was published by William Morrow in 1928? It assured Americans of two things they wanted to believe—that there was a place, albeit far away, where sexual freedom actually existed and even worked, and where family systems also functioned better than in America. Although the late 'twenties were the Jazz Age, memorialized in *The Great Gatsby*, the touted freedom of flappers and their male counterparts may not have been as great as its myth-makers made it seem. At the behest of William Morrow, whose expertise was in educational publications, Mead added two final chapters to her monograph about Samoa to make explicit the comparisons between America and Samoa that she had already signaled in lectures and conversation. These two chapters, entitled "Our Educational Problems" and "Education for Choice," suggest that the late 1920s were not as sexually liberated as F. Scott Fitzgerald's book depicts. Mead describes family life in America as a theater of coercion. She describes an American culture that offers so many choices—of religion, occupation, sexual behavior, values, and kinds of marriage—that the adolescent necessarily becomes confused. This confusion easily turns to resentment and rebellion when parents disallow choices other than their own. To reduce this tension, Mead advocates that we teach our children "how to think, not what to think" (*CAS* 246).

Mead was not the only observer of American family life in the late 'twenties and early 'thirties who found it oppressive. Joseph Folsom studied the same topic in 1934, and concluded that "all sexual behavior on the part of children is prevented by all means at the parents' disposal.... For the sake of prevention it has been usual to cultivate in the child, especially the girl, an attitude of horror or disgust toward all aspects of sex" (qtd. in Holmes 107). Willystine Goodsell's 1928 findings concur with Mead's when she states "That the home is not successfully meeting either the demands of society or the deepest needs of its members is evidenced by the prevalence of juvenile delinquency and crime, by outbursts of suicidal mania among youth, by the establishment and spread of child guidance clinics, juvenile courts and the probation system" (Holmes 108). Sociologists paint a picture of 1920s and '30s America that would be fertile ground for the imagination of tropical elsewheres, and Mead fed that imaginative need with her treatise on the *fa'a Samoa*.

Another reason for its popularity is that *Coming of Age* reads easily. Her writing style is remarkably clear, in part because she uses laymen's terms accessible to the non-anthropologist, in part because she adopts a novelistic style in which little dramas are replayed through the benevolent viewpoint of an omniscient narrator. The first chapter, "A Day in Samoa," sounds like a film script, establishing the *mise-en-scène* in idyllic terms: "As the dawn begins to fall among the soft brown roofs and the slender palm trees stand out against a colourless, gleaming sea, lovers slip home from trysts beneath the palm trees or in the shadow of beached canoes, that the light may find each sleeper in his appointed place" (14). Far from the tedious effort of understanding arcane kinship systems—anthropology's main turnoff—Mead's book offers a dramatization of the typical family group. She shows that what happens under the thatched roof of the sleeping house counterbalances what happens "under the palm trees," the Samoan euphemism for lovers' trysts. There is a strong family unit, regardless of easy adoptions; its strength comes from its fluid boundaries.

Of all islands, tribes, or peoples, why did Mead choose Samoa? First, she had a library-based familiarity with the general culture, having written her doctoral thesis on stability and change in Polynesia. In books, she had researched the tattoos, houses and canoes of Hawaiian, Maori, Samoan, Tahitian and Marquesan cultures, as indices of tradition and change. The prosaic and general flavor of this topic clearly reflects the difference between armchair anthropology and more exciting, focused fieldwork. Second, she had a challenge she wished to meet: she wanted her "own people," not some convenient group that had already been revealed to the American public. Her mentor, Dr. Franz Boas, wanted her to study the southwestern American Indians, since he did not think the place she had chosen, the isolated Tuamotu Island in the Marquesas, safe for a 98-pound, 5'2", 23-year-old American girl. But he also appreciated the strength of Mead's determination and her emotional nature; he decided it was better to send her closer to where she wanted to go. Edward Sapir, a distinguished linguist (and Mead's lover) questioned her emotional stability — was she emotionally equipped for isolated fieldwork in a tropical climate? Boas responded, "Of course I know that Margaret is high strung and emotional, but I also believe that nothing would depress her more than inability on account of her physical makeup and her mental characteristics to do the work she wants to do" (Lapsley 121). He understood her well; Mead's needs could not be ignored or denied. She had already learned how to get her way with male authorities by manipulating around her father's autocracy. In this instance, she appealed to Boas's cherished self-image as a liberal, tolerant, "modern man": "It was enough to accuse him obliquely of exercising inappropriate authority to have him draw back" (BW 140).

Mead settled on Ta'u, a small, isolated island in American Samoa that had had little in the way of destructive contact with colonizers. Its American Naval Dispensary with American staff, and a mail-and-supply boat running every three weeks ensured relatively good contact with the outside world. The problem Boas posed for his student was whether Samoan teenagers experienced the *Sturm und Drang* of adolescence so prevalent in western cultures. The more general theoretical question she and other Boasians were trying to answer was whether human nature is malleable or rigid, whether we are more "determined" by heredity or more shaped by culture. Her conclusions, after eight months in the field, affirmed the latter. Furthermore, if adolescence is not intrinsically a period of stress, then America might learn from Samoa how to change its own paradigm. The dry, academic approach of her thesis material on tattoos gave way not only to an engaging portrait of a happy culture, but more importantly, to the characteristic reformist ethos in her writing, the idea that anthropology has a strong social purpose. After all, the impetus for changing her major (from English to Psychology to Anthropology) was Ruth Benedict's dictum: "Professor Boas and I have nothing to offer but an opportunity to do work that matters" (Lapsley 69). The sense of moral mission was to characterize Mead's projects for the rest of her life.

How did she arrive at her conclusions about Samoan teenagers? Mead lived on the screened back porch of the American Navy Dispensary and conducted interviews of a group of about fifty girls between the ages of eight and twenty-one. Slight and young as she was, Mead's great advantage was that she appeared to be the same age as her informants, and thus was able to disarm them into making their "confessions." To further equalize relations, she pretended to be unmarried, though she had a husband back home. Since they attended American-style schools, Samoan girls were used to taking tests. To assure the appearance of legitimacy, Mead administered a few cognitive tests in a borrowed schoolhouse before questioning girls on their sexual experience. Although Boas warned her not to spend too much time on a full-scale ethnography, Mead found it necessary and possible to study the family background of each of her interviewees. She left her porch, stayed overnight in families' sleeping huts, and went on fishing expeditions with them. She was given *taupou* (ceremonial virgin princess) status, and learned the ritual dance. One of the more famous photographs shows her rigged out in native dress, posing flank to flank with her "best friend."

Being taken under the wing of the High Chief gained her access to important information, but as a woman, Mead was not allowed to attend regular town meetings. Nor were women traditionally the carriers of genealogical and legendary information (this was the bailiwick of men), so Mead's female friends would not have relayed this important kinship and historical background. Mead does not deny that her picture of Samoan life was skewed to a female, adolescent viewpoint, since the bulk of her information came from the fifty girls in her test sample. More specifically, of the twenty-five girls between the ages of fourteen or fifteen to nineteen or twenty, thirteen reported no heterosexual experience, producing about fifty percent virginity among the adolescent girls of her study. Although this statistic may not appear to bear out her hypothesis that premarital sex was less

Margaret Mead in Samoan dress with her friend and informant, Fa'amotu, c. 1925. Mead attained a high degree of rapport with her female informants, despite the usual reticence, in girls, to speak of sexual matters. Part of the reason for her success, besides a willingness to dress and to dance ike them, may have been the appearance of being single, since her first husband was living in New York. Courtesy Institute for Intercultural Studies, Inc., New York.

frowned upon in Samoa than in America, it must be fit into the larger picture of the culture.

For Mead discovered other practices at variance with American family and dating life, and these painted a full panorama of the life of relative ease and harmony which so attracted her readers. These practices included ease of adoption, ease of divorce, ease of elopement, and the acceptance of illegitimate children. Key to the easiness of these institutions is what she calls the "absence of specialized feeling," meaning that affections are light, and diffused among many objects. Thus, neither the parent nor the love interest is cathected in the way he or she would be in western societies. Samoan children become comfortable with their bodies and sexuality at a young age; they are called upon to dance in public ceremonies. Even the *moetotolo*, the surreptitious night rapist, is softened in the way that Mead explains him. Good-looking but frustrated in his attempts to achieve the intercourse refused him by the object of his affections by day, he visits the girl in her family quarters under cover of darkness. Since the family sleeps together, the girl has the choice to rouse them to her defense, or quietly to accept the advances of the surprise visitor.

Other than the ease of premarital sexual relations, and the lack of jealousy surrounding the sexual relationship, the ease of adoption made the most profound effect on Mead, especially in contrast to her own culture. In Samoa, when children found their parents' discipline or temperaments too annoying, they merely moved to the home of another relative. Households consisted of extended families of persons related to the head of family (*matai*) by blood, marriage or adoption. Lines of authority in each household were clear. Age was the classifying principle, and anyone of a greater age had automatic authority over anyone of a lesser age. Although this system appears potentially unstable (what if a child fails to accept any authority whatsoever?), the reason it functions is the deterrence value of threatened exile. If a child uses up all of his available options, he can be exiled to another village. In such a traditional culture, this is a particularly unattractive alternative.

Another stabilizing feature is temperament. Mead's interviews and observation proved that the most highly valued virtue was kindness, and the emphasis was on cooperation rather than independence, so delinquents and truants were the rare exception. Mead was fascinated by the adoption solution to the problems of the nuclear family, and would propose in a 1943 essay ("The Family in the Future") that American families should foster out their discontented offspring. However, to find foster parents for unsuitably matched parents and children in a heterogeneous, industrialized country would be cumbersome and probably impossible on a system-wide basis, though individuals might attempt it. Her own lifestyle was an attempt to avoid the volatility and isolation of exclusive intimate relationships by creating her own extended family. She nurtured friendships, but perhaps more importantly, fostered a system of dependence whereby others needed her for their careers or other types of support.

As in her accounts of her own life, Mead minimized the conflicts in her description of Samoan life, and summarized it as an ideal conjunction of pagan and Christian values and beliefs. Acknowledging that Samoa had been Christianized since the 1840s, Mead nevertheless asserted that Samoans knew how to adopt the good and reject the bad or incongruous in the missionaries' teachings. In Appendix III, "Samoan Civilisation As It Is Today," she praises both the colonial administrators, for "benevolent non-interference," and the Samoans themselves for being flexible enough to blend European and primitive ways. Samoan law was not always so flexible; it required "white influence" to soften Samoan law and order. The most dramatic example of Samoans' historic inflexibility is the beating — occasionally to death — of the ceremonial virgin if her hymen does not bleed from a public "deflowering" (either groom or chief inserts fingers into her vagina). This mix of the horrific and the serene (in the Samoas of yesterday and today) possibly accounts for the popularity of the book.

Although it is pleasant to tell a story of unruffled success and satisfaction, the recent availability of the papers of many of Mead's friends, family and colleagues has turned up another side to anthropology's Giant. Historian Lois Banner (*Intertwined Lives: Margaret Mead, Ruth Benedict, and Their Circle*, 2003), had access to Mead's mother's diary, Benedict's sister's papers, and those of Geoffrey Gorer (a close collaborator and homosexual friend whom Mead considered marrying), Robert Lowie (one of Boas's disciples), and Léonie Adams (one of the Ash Can Cats— Mead's circle of friends from her Barnard days). Banner tells a nuanced story of Mead's rise to fame, including her insecurities and depressions during her first fieldwork. In Samoa, she "could hardly cope, lacked confidence in herself, and obsessed over her relationship with Sapir" [the linguist who had asked Margaret to marry him] (234). Less generously, Derek Freeman analyzes Mead's letters from Samoa with an eye for how her insecurities produced or revealed faulty methodology and misleading conclusions. After three and a half months, Margaret complained to Boas that she had not found anything interesting or tangible about adolescent sexual behavior. After another three months, she wrote to Benedict that she was sick of talking about sex with her adolescent girls all day long (Banner 239). She confessed that she cried all day long, hated the heat and humidity, loathed the bugs, had neuritis in her arm and, with her "inferiority complex," was convinced she was going to fail in her project. Her black humor returns with the line, "I'm just going to give up and get a job taking change in a subway" (Banner 240).

One of the paradoxes about Mead doing fieldwork in isolated South Seas islands is that she loves cities, especially New York, which she made her life-long home as soon as she escaped from her nuclear family. Conversely, she finds rural life boring. As her friend Martha Ullman West noted, "She hated the country.... She conceded that flowers were pretty ... but plants and animals, as far as she was concerned, were dull. She was absolutely an urban person. What she cared about most were people, in the largest possible doses" (Howard 311). Even though she loved people, Mead experienced the common ethnographer's dilemma: she wanted extensive contact so that she could understand the culture; on the other hand, locals with time on their hands thought nothing of giving her constant companionship. The concept of a private abode was alien to them. Nor was it proper for a woman to go about the village unaccompanied. In her little porch-bedroom, Margaret didn't get enough solitude; the girls quite naturally regarded her as a novelty, and visited her at all times of day. "I never get any time alone," she complained to Benedict (Banner 239). Yet she was very lonely. While in Samoa, she vowed never to do fieldwork alone again.

Mead's daughter, Mary Catherine Bateson, describes her mother's loneliness in Samoa as more than personal; it was cultural isolation writ large. The doctors in the dispensary were too puritanical and racist to be fun, but constant exposure to the Samoans did not provide Margaret with the kind of social comfort she needed either. Fieldwork is a trying mental exercise; confronted by difference, one must consistently prepare the mind to meet it. "Instead of culture shock one should probably speak of this kind of experience in the field as culture stress, continuing and cumulative. It is a stress that leads to a questioning of the very self," wrote her daughter (Bateson 130). Mead later corroborated this in describing feelings she had while studying the Arapesh people of northern New Guinea. When her husband left the village for excursions she could not join because of physical limitations, she "had too much time to think" (BW 199). Mead loved to think, but did not like to do it alone. Colin Frank, who grew up in a cooperative household with Mead and her daughter, comments upon Mead's remarkably unreflective temperament, as well as her utter disinclination to undergo psychoanalysis, even though she glibly recommended it to friends and strangers. Martha Ullman, a friend of Mead's daughter and Colin, reports that Mead said that some people "need to bury things rather than drag them out," and that she was one of these (Howard 253). Mead never did like time alone. When she set foot on the ocean liner that would take her to Samoa, she realized she had never, at twenty-three, spent a night alone in her life.

With the recent availability of correspondence, the darker side of the Samoan trip is thus revealed. Mead felt particularly insecure about her work methods. She felt she had to teach herself the methodology of her profession because Boas only talked to her for "half an hour" about how to do field work.[1] She was also concerned about her love life. She had already begun to feel estranged from her husband, Luther Cressman, after a year in New York. Her first extramarital love affair, with Edward Sapir, had ended badly. He expressed in letters the anger of a jilted suitor, one whom she considered her intellectual superior (and thus extremely compelling). Edward Sapir was her first exposure to the perils of mixing business with pleasure. When she rejected him, he damaged her reputation, telling Boas's principle disciples, Robert Lowie, Alfred Kroeber, Paul Radin, Alexander

Goldenweiser, and even Cambridge anthropologist Bronislaw Malinowski, that she was a prostitute and a liar. A traditionalist, he did not see how she could love him sexually, and then refuse to marry him. In Samoa, she was also in turmoil about her own sexual orientation, having begun an intimate relationship with her mentor, Ruth Benedict, two years earlier. Who should she choose for a monogamous relationship when she returned to New York? Could either Luther or Ruth force her into monogamy when she did not believe in it? Her return voyage answered these questions with a third alternative. On board ship she met an intense, manly psychology student from New Zealand with the exotic name of Reo Fortune, whom she was to marry and convert to anthropology. Although there were sexual chemistry and shared interests, this was a volatile mixture of personalities. Coming from a traditional, provincial background, Reo was not willing or able to take the back seat to Mead, who was further advanced in the field than he. They would do their next two field trips together, in Manus (Admiralty Islands) and Sepik (New Guinea), but the stresses of temperament and difficult field conditions would soon divide them.

But at the time of her return from Samoa, these battles were on the distant horizon. Besides a new husband, another good surprise awaited: Mead's field work, written up as the *Coming of Age* monograph and a separate, more technical ethnography, cemented her good standing with Boas. Apart from the comment that she failed to differentiate romantic from sexual love, Boas made no criticisms of her monograph on the Samoans, and even wrote a highly complimentary Foreword, a favor he had not extended to the many male graduate students he had mentored. "We feel grateful to Miss Mead for having undertaken to identify herself so completely with Samoan youth that she gives us a lucid and clear picture of the joys and difficulties encountered by the young individual in a culture so entirely different from our own," he said (*CAS* Foreword). In addition, America's renowned sexologist Havelock Ellis wrote praise that became part of the cover design of her book. Even Malinowski rose above Sapir's sexual gossip and proclaimed it "an absolutely first-rate piece of descriptive anthropology ... an outstanding achievement" (Lapsley 194). She received five thousand dollars in royalties the first year. Within a year and a half of her return from her first challenging fieldwork, she was gracefully launched into one of the most publicized careers in America.

Killing the Giant: The Mead/Freeman Controversy

Even after her death in 1978, Mead's life continued to be held up to the public eye. In 1983, Derek Freeman culminated a career's worth of research into Mead's possible errors in Samoa with his monograph, *Margaret Mead and Samoa: The Making and Unmaking of an Anthropological Myth*. Among other things, he has been accused of lacking the courage to publish the book while Mead was still alive. Freeman was a New Zealand anthropologist working at the Australian National University. The "myth" of his title is the proposition of absolute cultural determinism, and he is the self-proclaimed "unmaker" of that myth. It was a common myth in anthropology that the ethnographer in the field was a larger-than-life hero, who dauntlessly faced isolation, savagery, disease, possible murder, and cannibalism, in the interests of scholarship and civilization. But the myth of the anthropologist as dragon-slayer was a new one.[2] Freeman was trying to prove that "the real enemy was within"—within his own field, that is. He confessed to biographer Jane Howard that he believed his book would cause Mead's reputation to "do a 32," that is, plummet through space at the rate of speed of falling bodies (436). Perhaps misunderstanding that in America, fame is built on more than quality of work, Freeman erroneously believed that Mead would be forgotten if he discredited her. His modus operandi was this: if Mead's Samoan research was assumed to prove that culture, not biology, made adolescence either easy or difficult, then showing her research to be faulty would in turn disprove the validity of cultural determinism. The main flaw in this argument is that Boas, Benedict, Mead and other founders of cultural anthropology were not the absolutists that Freeman depicts. They were never cultural determinists, says Lowell Holmes, author of *Quest for the Real Samoa*. Freeman distorted their position—which was that both biological and cultural factors shape personality. He overstated the case in order to provide a motive for

Margaret to dissemble. In its quest for dark motives, Freeman's "unmaking" of Mead shares many components of a good — or perhaps bad — detective story. As Freeman tells it, Boas nursed a personal vendetta against G. Stanley Hall, President of Clark University, from which Boas had resigned on a principle of academic freedom. Hall's book on adolescence propounded a straight biological theory: adolescence is stormy because of hormones. In Freeman's view, Mead was a pawn in Boas's ego contest with Hall. If she could prove that Samoan adolescence was stress-free, then Boas would be vindicated. Freeman asserts that she ignored evidence contrary to their hypothesis in order to please her mentor. Fortunately, the Boas-Mead letters have been carefully analyzed by other anthropologists, including Martin Orans, Lowell Holmes, Paul Shankman, Stephen Murray and James Cote, who all arrive at the opposite conclusion from Freeman's. They find that Mead was not Boas's pawn, and that she was convinced of the truth of her findings.

Lowell Holmes has done more than any other anthropologist to clarify the Freeman-Mead controversy. He studied anthropology at Northwestern University in 1954. His advisor recommended he write a thesis on Mead's Samoan research. The first methodological restudy ever undertaken in anthropology, Holmes's project was not intended to discredit the Giant, but rather to understand more about the process of fieldwork itself. As Holmes states (my italics):

> A major consideration throughout the study was whether differences between my findings and Mead's were due to *cultural change or other personal or methodological factors*. Because of this problem, I had to document the culture for the periods of (1) the mid-nineteenth century, a period of prolific ethnographic writing by missionaries; (2) the 1920–30 period, when Samoa was studied by Mead, Peter Buck, N. A. Rowe, and Felix Keesing; and (3) the contemporary scene as I observed it in 1954.... And, in order to evaluate her statement that it was easier to come of age in Samoa than in the United States, I also found it necessary to investigate what it was like to be a teenager in America in 1925. (14)

Holmes returned to the same village in which Mead had lived, interviewed the same people, and discounted many differences in findings because of the historical time difference, the difference in his gender, and in the age of the informants between Margaret's visit and his. Freeman made no such careful controls. He did his own fieldwork, not in American, but in Western Samoa, which was a territory of New Zealand until 1962. Western Samoa was very different from American Samoa; the islands were larger, the colonial attitude more apparent, and the ills of contact more severe. The growth of cities had fostered crime and alcohol abuse. As Felix and Marie Keesing noted, New Zealanders kept a greater distance from natives than Americans did: "most New Zealanders avoid easy comradely behavior with Samoans both in and out of government and keep a considerable social distance" (193). The Keesings observed that Americans have a less colonial mentality and treat natives more like colleagues, and that a different colonizer created different types of colonial relations. Freeman also ignored the difference in time. Mead was in Samoa in 1925, Freeman in 1942, 1967, and 1981. Because of the growth of cities, industry, and crime during that period, culture had considerably changed. Finally, he failed to account for his gender. He assumed that because of his male authority, his interviews with adolescent girls about their sexual experience were free of the joking, hoaxing, and duping of which he claimed Mead was the victim.

Seeking further objectivity through comparison, Holmes consulted psychometric assessments made by fifteen other scholars of Samoa. Conscious of the great influence of the ethnographer's temperament on the outcome of his or her ethnography, anthropology in Holmes's day had begun to address these problems directly. As an attempted counterbalance to the subjectivity of observation, Holmes adopted various diagnostic tests that may have been more objective, including "folklore and literature analysis; projective, verbal, and non-verbal testing; controlled laboratory observation; personality inventories; value schedule analysis, and others" (Holmes 130). The result: "a profile of Samoans very much at odds with that put forward by Derek Freeman" and "strongly supportive of Margaret Mead's conceptualizations" (136). Where Freeman found in Samoa a population of aggressive, possessive, maladjusted, selfish, and sexually repressed individuals, Mead found a population of harmonious, peaceful, cooperative, well-adjusted, generous and sexually expressive communitarians.

Although Holmes found the Samoans different from what Mead described in 1928, he characterizes these differences as interpretive rather than factual, as a difference in their perceptions of the ethos of Samoan culture. Donald Campbell highlights the subjectivity of all judgments of ethos. The observer's ethos inevitably intrudes: "It is quite conceivable that there are some aspects of culture, including its overall pattern or ethos, that are so abstract or indirectly inferred that intersubjective verifiability is lost" (444). In *Growth and Culture* (1951), Mead comments that "our own emotional response" to a culture is the most "perfect instrument of understanding" (300). Her own strong ethos is what made *Coming of Age in Samoa* such a success. Clearly, the solution is not for the ethnographer to try to curtail his or her emotional response; instinctive as it is, this would be impossible. The solution, according to Campbell, is to open up the field, to welcome observing anthropologists from different cultures for a "formal combinatorial analysis" (444). This is exactly what Holmes did.

Today, generalizations about national character inevitably raise the red flag of "nationalistic prejudice" (as Freeman has characterized the backlash against him). On the contrary, in the 1950s, statements about national character were not considered amiss (in fact, Mead and a group of anthropologists undertook several studies of national character in the 'fifties). The Keesings were thus "allowed" to comment on the possibility of a national-character bias in Freeman: "The New Zealander ... tends to be critical of the image which Samoan traits convey to him, e.g., the stress on hierarchy, the flair for dramatic display, ceremoniousness, and circumlocution, the emphasis on group-responsibility, with externalized shame-sanction, and suppressive (group-regulating) controls" (305). A number of other critics have expressed their view that Freeman's work on Samoa reveals more about him than about Samoans (his own moodiness and competitiveness, among other things). Likewise, Mead's work projects her own temperament, as well as her wishes, upon the Samoans, particularly her vaunted optimism and her wish for gender and racial equality and sexual permissiveness. Between these two poles—Freeman's negative and Mead's positive—a mediator has come. Holmes's 1954 restudy, as well as his 1987 book-length response to the Freeman controversy, provides reasonable explanations for the discrepancies between Mead's and Freeman's interpretations, while at the same time highlighting the perils of the profession itself, given its subjective nature.

Freeman's motivations for debunking Margaret Mead are clearly complex and multiple. Serge Tcherkézoff, author of *Le mythe occidental de la sexualité polynésienne: Margaret Mead, Derek Freeman et Samoa* (2001), claims that Freeman's animus against Mead was actually driven by a deep love and patronage of Samoans. Since the 1960s, several Samoan intellectuals have commented that Margaret Mead's book did their people a disservice in depicting them as libertines. "Long standing pre-missionary Samoan values require that any reference to the sexual act be distanced from the Samoan social proprieties," writes Tcherkézoff (435). Mead's depiction of their society violated this "distance," and Samoans who traveled abroad had to face this libertine stereotype. Albert Wendt, distinguished Samoan novelist, poet, and scholar, concurs with Tcherkézoff when he writes, "He [Freeman] has a deep love and respect for us. This I think helps to explain his almost obsessive quest to correct what he deems was the wrong Margaret Mead did to us. Perhaps he has not felt at home in his own society, and in understanding us hoped he would find a people to belong to, to champion, to be needed by. The condition of the outsider is one I know well" (12).

This interpretation of motives softens our view of the Giant Killer, yet Tcherkézoff offers it largely out of respect for the recently dead. The occasion for his article is to clear his name of defamatory charges that Freeman made before death—charges that Tcherkezoff, in a fit of "Gallic passion,"[3] flagrantly fabricated a story about Freeman to "demonise" him (Tcherkezoff 434). What did Tcherkezoff allegedly fabricate? He said that in 1987, Freeman had forced one of Mead's informants to confess before a video camera that what she had told Margaret about having more than one boyfriend was a lie. According to Tcherkézoff, he shamed her into tears, not because she had lied, but because her lie originated the widespread stereotype of Samoans as libertines. Tcherkézoff believes that Freeman made her feel that her playful lie single-handedly cost the Samoans their reputation in the international eye. Calling Freeman "unethical," he claimed that this act should put him "out of the anthropological community" (434). This is the insult that gave rise to Freeman's angry rejoinder. Only after Freeman's death did Tcherkézoff gain a better understanding of his motives for lashing out. Freeman was so attached to "his people" (the Samoans) that Tcherkézoff's charge of insensitivity to them deeply offended him. This interpretation corroborates Wendt's theory that Freeman identified with Samoans and was protective of their good name.

The Mead/Freeman controversy gives insight into the overt rivalry in this academic field. Anthropologists feel territorial about their specializations, affectionately calling them "my people." They also become attached to their own interpretations of these populations. Mead's famous toughness was an asset in such an arena, although her life-long pursuit of recognition outside the field was perhaps her best defense against the competition within it. For Mead's was very much a public, rather than an academic, career. Her promotions within the museum were very slow in coming — sixteen years to rise from assistant to associate curator, twenty-two more years to become full curator. But she did not contest the process, as many women might do today. She regarded museums as safer than academia, less likely to fire people for arbitrary reasons. When she accepted the job at age twenty-four she vowed that she would stay as long as she could climb the steps to her top-floor tower. Her job lasted a lifetime; it was convenient and easy. The museum gave the extensive leaves of absence she needed for fieldwork and public lectures, in return for a minimal amount of curatorial work. If Margaret Mead was not a typical academic, nor was she a classic museum person.[4] Collecting was not her love; she preferred people to objects. First and foremost, Mead was a public figure, one who wanted and needed the public eye. Mead learned oratorical skills early; her father taught her to look her audience in the eye (both literally and figuratively). She sought out as much of the world's limelight as she could possibly fit in her schedule between her two jobs as curator at the American Museum of Natural History and as adjunct professor at Columbia University (she never got on tenure track there). The jobs were secondary to her real career as public figure. Patricia Grinager, a former student and life-long friend to Mead, describes how Mead developed her "real career." It is worth quoting at length for the insight this passage gives into Mead's capacity for self-promotion.

> Her career as a formal speaker began almost as soon as she returned from Samoa. Along with churning out articles about her fieldwork for such periodicals as *The Delineator*, *Parents' Magazine*, *The Nation*, and *Natural History*, she addressed the membership of her mother's numerous ladies' clubs. She described the daily routines of South Sea Islanders whose existence barely resembled the everyday lives of urban or suburban Philadelphians. Her topics piqued their interest. Titles such as "South Sea Hints on Bringing Up Children," "Living with the Natives of Melanesia," "Water Babies of the South Seas," and "Going Native for Science" titillated the curious to invite her again.
>
> As her writings became known, invitations to speak arrived from communities far beyond Philadelphia and New York City. Conference planners representing churches, clubs, various special interest organizations, universities, and professional societies requested her presence. She obliged as many of them as time permitted and pursued her lecture trail experiences with the energy of a political campaigner. She frequently declaimed before two or more entirely different interest groups between breakfast and bedtime. She became as familiar to college-town America as had nineteenth-century headliners for the Chautauqua in her grandmother's day [52].

If we consider how difficult it is to make one's mark in anthropology or in any academic field, perhaps the constellation of reasons for Freeman, Sapir, Malinowski, and other male anthropologists to want to kill the giant becomes clear. The list of things they would have envied about Mead is long, starting with her early fame, at twenty-seven. She was popular in countless venues. Unlike many academics, she could write well and without hesitation. An extrovert, she had a knack for making friends. She could lecture fluently without notes, keying her subject matter to the responses of her audience. Her university lectures were filled to overflowing. After *Coming of Age in Samoa*, she found publishers for her work with ease, and proceeded to write, co-write or edit thirty-nine books, 1,397 articles, numerous reviews, and other short pieces. She received huge sums on the lecture circuit (which she made over to the Institute for Intercultural

Studies and graduate fellowships). Renowned psychiatrists valued her as an expert on human behavior; for twenty-one years she was Visiting Professor of Anthropology at the University of Cincinnati College of Medicine. Officials of many church denominations welcomed her into their folds and were enlightened by her. She wrote familiarly to President Jimmy Carter, advising him how to proceed with his plans for social reform. She knew how to network and to delegate. She received twenty-eight honorary degrees, forty awards, and membership in the American Academy of Arts and Letters. She was President of the American Anthropological Association, the Society for General Systems Research, the Anthropological Film Institute, the World Society for Ekistics, the Scientists' Institute for Public Information, the Society for Applied Anthropology, and the American Association for the Advancement of Science.

These organizations are so male-dominated that perhaps women had reason to envy her as well. She resolved that potential problem by having an immense appreciation for both female friendship and female collaboration. Uncannily, she knew how to cajole, command, reprimand and demand women to do their best work for her, usually without losing their loyalty in the process. In return, she launched their careers. She was a powerful role model for women, even though, as she grew older and more confident, she could be quite a tyrant. Howard believes that friends excused her selfishness because they could see the vulnerability inside the giant (xv). I think they could also see her greatness, and that they allowed their appreciation of her messianic spirit to outweigh the personal offensiveness of her style. In my view, she had the kind of intelligence that people most admire — smart enough to understand theory, practical and moral enough to apply it for social betterment, and social enough to express her vision in terms that everyone could understand. The combination is so rare in academia that no comparable intellect comes to mind.

Mother of the World,[5] *Mother of One Child*

I am fascinated by Margaret Mead's capacity for a vibrant social life, which was centered around her work interests in the first half of her life, and a network of professional colleagues and acquaintances in the second half. Although the second half might be considered a sad story, insofar as she disported herself like a diva and lost her down-home quality, in some ways she did achieve the steadfast coterie that D. H. Lawrence envisioned and could never accumulate. Not a social life in the partying or flirtation sense, Mead's domestic and social groupings, from the time of the Ash Can Cats onwards, were characterized by their shared work ethos and ideological commitment to either rebellion or reform. Especially during college, women who pursued careers as well as male and female friendships were considered rebellious from the mainstream viewpoint that college was a place to find a husband and settle down. The shared New York domiciles with first and second husbands Luther Cressman and Reo Fortune were Grand Central Station for Margaret's friends. There these young women found a home: they could eat, talk, sleep over, and receive consolation for their tattered love lives. Yet many of the visitors were also involved in serious work of an intellectual or artistic cast. Poet Léonie Adams was in the inner circle, and through her, more famous poets such as Edna St. Vincent Millay, Amy Lowell, and Louise Bogan were introduced. Though Margaret considered him "a decadent little egoist," critic Edmund Wilson was in, and had tempestuous affairs with Adams and St. Vincent Millay, and a disastrous marriage to Mary McCarthy (Banner 222). Louise Rosenblatt was a recruit to the Cats, and later became a literature professor at Brooklyn College and co-founder of reader-response theory. Ruth Benedict, who made her mark in anthropology for her theory of culture, personality, and deviance, was Mead's closest friend, collaborator and soul mate, although fourteen years her senior.

These early apartments, the epicenters of various intellectual circles, provided a model for Mead's later Perry Street experiment in cooperative living. In 1939, Margaret returned from fieldwork with her third husband, Gregory Bateson, looking for a place to live. Pregnant with her first and only child (she was thirty-eight), she soon entered into a living arrangement at Perry Street, Greenwich Village, that was to stand as her most successful experiment in cooperative living, at least from her point of view. The Lawrence Franks owned a five-story brownstone. At a social sciences seminar in Hanover, New Hampshire, Larry met Mead and found her ideologically compatible; he invited her and her family to move in. Larry was a funding administrator for the Rockefeller Foundation; he funded child-study centers and interdisciplinary social sciences conferences. He had a "near religious zeal for interdisciplinary enterprise," an emotional quality his sons found somewhat disturbing in later life (Banner 348). Mary Frank, Larry's young second wife, did not seem to mind taking care of Margaret's new-born while Margaret worked, even though she had three of Larry's children from a previous marriage, and two of their own. Mary and Larry's sons, Colin and Kevin, had a somewhat different view of the utopian experiment. So did Mead's daughter Cathy, who later expressed regret that her mother had left her with the Franks for days at a time while working in Washington, D. C. Colin Frank describes Margaret as intimidating and demanding. As an adult, he tries to appreciate the unique upbringing he shared with Cathy, while he also questions the cost to his mental health.

Mead and Gregory Bateson, her third husband and father of her daughter, Mary Catherine Bateson, c. 1934. They fell in love in New Guinea where they worked on neighboring tribes. Although Mead was then married to Reo Fortune, the three collaborated on innovative anthropological theories. Reprinted by permission of Sll/sterling Lord Literistic, Inc. Copyright by Estate of Jane Howard.

Mead's last non-nuclear, alternative family spanned over two decades. A few years after her divorce from Gregory Bateson, Mead moved into the spacious Waverly Place, Chelsea District apartment of Rhoda Métraux, also an anthropologist. Rhoda had earned a Ph.D. from Yale, worked as an assistant for Mead at the Committee on Food Habits in D.C., and then as an assistant for Bateson in the Office of Strategic Services in 1943. After a short affair with Bateson, she became Margaret's live-in companion for the rest of Mead's life. As Mead spent more and more time on the road after her divorce, her New York home became less of a social center.

Less scholarship has been done on the Mead/Métraux relationship than any other of Margaret's marriages and intimate relationships, but what emerges is a pattern not unlike that which Margaret shared with Marie Eichelberger. Marie was a friend from student days who "worshipped" both Mead and Benedict. Her chosen career — social work — did not end when she came home at night. From college, when Marie mended Margaret's clothes, cleaned her room, and did her hair and nails, she graduated to more professional and extensive care-taking of her hero, including paying her bills and taxes, caring for her daughter, and doing other services typical of a personal secretary or servant. Since she was an anthropologist with her own career, however, Métraux's chores for Mead were professional as well as personal. She co-wrote or drafted articles for Margaret's signature, screened her calls, accompanied her to conferences and to the field, and generally held her overscheduled life together with a human touch. Two scholarly biographies have been written about the friendship between Mead and Benedict in the past five years, but none on Métraux. Once Métraux's papers are available, there should be another book on this significant relationship. Until that time, sources appear to be protecting her privacy, especially in the aftermath of the Freeman controversy.

How might the relationship be viewed as controversial, besides the fact that it was homosexual? Unlike the relationship with Benedict — that began as mentor and protégé, but quickly became one of equals— the one with Rhoda was never on an equal footing. Mead had already made her name by the time she met Métraux. Detractors could easily see in this relationship an opportunity to brand Margaret as a people-user. Grinager and Howard provide some sense of the relationship's intimacy when they describe Métraux's jealousy. In her seventies, Mead developed a strong intimacy with Jean Houston, a brilliant philosopher-poet the age of her own daughter, whom she promptly "adopted." She also had a healing physical and emotional relationship with a psychic, Carmen de Barraza. Métraux was jealous, and their tensions drove Mead out of their apartment for several months when she was ill with cancer. When anthropologist Patricia Grinager and her colleague Dana Raphael collaborated in a project documenting Mead's minute-by-minute activities at the

American Anthropological Association meeting in Mexico City in 1974, Rhoda also appeared jealous and protective of Mead. From Grinager's account, Rhoda's job seemed to be one of maximizing professional contacts for Mead, while minimizing personal conflicts—damage control, as it has come to be called. Was Métraux a glorified P.R. agent for her best friend; was this another case of Eichelbergerian hero worship, only at a higher professional level? The next biography or memoir will certainly address the question.

The four domestic arrangements just described — Ash Can Cats, first two marriages, Perry Street, and Waverly Place — have a few things in common. In all of them, Margaret is, or perceives herself as being, the boss. In fact, when the Ash Can Cats gently reminded her that they had existed before she transferred to Barnard for her sophomore year, she was nonplussed and denied it. In her first married household, it was lucky for Mead that Cressman enjoyed the company of her female friends and tolerated Margaret's frequent absences, because at times it certainly seemed that her own social life was more important than their marriage. It was more difficult to orchestrate such a communal life with her second husband, Reo Fortune. He came from a more traditional family, and his father beat his mother. In addition, his class values dictated that he and Margaret shouldn't do housework; she describes in her autobiography how she had to do it surreptitiously on days when their housemaid was absent. The marriage did not last; the violence in the family history reappeared when Reo struck Margaret and she miscarried their child. The professional competition between the two was not ultimately worth the collaboration. As for the Perry Street household, Colin Frank provides accounts of Mead's dominant role, and Mead's needs were clearly predominant in the Métraux/Mead ménage. The other thing that all of these homes have in common is that they are alternative, non-traditional, non-nuclear-family arrangements. For Mead, the family must be communal in order to be safe, to blow off the steam of too-intense intimacy. If Samoa was her model, then Mead became the *matai* or head of every household she constructed. She creatively endeavored to shape families that were flexible yet dependable, open-minded yet ideological (egalitarian, progressive, liberal) — homes in which individuals could blossom while still being members of a community. Perry Street, and its summer-time extension in New Hampshire, were the best living arrangements for Margaret.

On the other hand, an eight-by-eight foot "mosquito room" in the wilds of the Sepik River region of New Guinea was the home for a unique and timely constellation of anthropological minds that exploded with ideas, emotions, libido, and dreams. I am drawn to this short-lived home because of the illustration it provides of the power of a third person to break the binary of two and inspire good work. It was the number "four," however, that guided their discussions, as they constructed their theory of "Squares." Their square was made of four personality types into which all humanity could be fit, building on both Jungian personality types and Benedict's "patterns of culture." I refer to the first four months of 1933, when Reo and Margaret studied the Tchambuli peoples of northern New Guinea, while Gregory Bateson commuted from his base with the Iatmul people, a few hours away. Charmed by both Margaret and Reo, and feeling isolated and lonely in Iatmul, Gregory frequently visited their home in its panoramic setting by an ebony-colored lake dotted with lotus flowers. Despite the prettiness of the setting, the human conditions were not very comfortable. The mosquito room was hot, humid, small, and buggy despite the netting; each of the three ethnographers contracted malaria and was occasionally delirious; Margaret and Gregory experienced an immediate attraction; Reo knew it, and was jealous. There was a constant danger of raids from neighboring tribes. However, the isolation and intensity of the human combination resulted in mental breakthroughs and profound bonds. It was here that Margaret found an answer to the philosophical question that was to unite the three very different New Guinea cultures she had so far studied.

The question all three anthropologists were asking was this: Are masculine and feminine character traits more determined by biology or by culture? The Tchambuli, in conjunction with the Arapesh and the Mundugumor, provided the ethnographical clue. But it was the sexual attractions and tensions in that small room that personalized the theory and drove it home, as the three intellectuals tried to assign their temperaments to one of the four squares and determine which temperament went best with which other. The Arapesh were altogether gentle; both males and females nurtured children and cooperated in the smooth flow of community activity (Reo found this boring; Margaret, pleasant). The Mundugumor of both sexes were aggressive, competitive and cruel. Because of a complicated system of inheritance, only children of one's opposite sex were valued, so men would kill male infants and women would kill female babies. Young children were terrorized by the threat of violence. The Tchambuli, on the other hand, displayed strong gender divisions, but in reverse of western culture. Women were in charge of food gathering and survival activities; men were marginalized and exhibited the petty, emotional characteristics of those whose importance is devalued — the negative side of femininity in our own culture. In Gregory's people, the Iatmul, gender roles were the same as traditional western ones, with strong, aggressive men and passive, nurturing women. Their dance rituals involved cross-dressing and parody of the sexual roles, which he took as a celebration and affirmation of them.

From these four examples of cultural gender "assignments," the anthropological trio deduced that, apart from reproductive capacities, gender roles are not primarily biological, but cultural. Individual temperament is predicated on culture, which determines how the sexes behave. Mary Catherine Bateson sums up her mother's revelation: "Gender — biologically given — supplies a component of temperament but the imprint of culture is strong enough to reshape a given temperament to make most individuals fit cultural expectations, including cultural ideas of masculinity or femininity" (131). The mosquito hut revelation stands as a prime example of professional collaboration while at the same time demonstrating Mead's belief that breaking the nuclear unit — while painful — results in more fruitful relationships. Both work and home, in Mead's view, must be shared.

Another positive example of Mead's family-making involves academic conferences. They are an institution that some scholars dread and others love. They can be competitive, intimidating, and cliquish. Graduate students are encouraged to attend, not only to practice speaking before a crowd of peers, but also to network for that increasingly rare attainment, a tenure-track job. Mead loved them. To her, conferences were scintillating: "For the exchange of ideas, Mead knew of no sublimer forum than the conference. Other eyes might glaze over at the prospect of days in stuffy rooms with sealed windows, of plastic badges, endless panel discussions: Mead's did not. Conferences, for her, seemed a highblown species of that splendid ritual the family reunion" (Howard 367). Mead of course did not approach the conferences like a graduate student. In discussions, she took the lead, opening topics up to point to more universal concerns. Conferences were opportunities to teach people "how to act as whole individuals, using all their senses as they seldom do in the narrower, more specialized contexts of professional and academic life" (368). Grinager's accounts of accompanying Mead to conferences are at times hilarious, as she shows how avid a networker her mentor is—quite the mingler. At the American Anthropological Association (AAA) meeting in Mexico City, she pumped two tall Afrikaners for insights into apartheid, to prepare for her upcoming lecture in the Netherlands; at the same time, her eyes "darted from face to face like a lost traveler in Grand Central Station" (171). No slacker at the hotel bar, or butterfly at the ubiquitous parties and "confabs" in people's rooms, Mead attended every possible session. She ordered Grinager to collect reprints of 108 of the presentations, and would try to read every paper before it was presented so she could say "intelligent things" (Grinager 168). "You can learn more in a concentrated morning listening to people read papers on their work than you can in reading for a month of Sundays," she advised Grinager (176). Unlike journal articles, conference research is up to date; it lets you know who is up and coming in the field. Conferences are that rare mix of scholarship and politics, a mingling of private and public realms.

Grinager's blow-by-blow account of the Mexico meeting, although written with Mead's permission, is both hilarious and sad. It reveals, as does the rest of her book, *Uncommon Lives*, a less savory side to Mead's knack for networks, showing that however professional its intention, it is also a manic avoidance of being alone. From Ruth Benedict's death in 1948 and Gregory Bateson's departure from their marriage at approximately the same time, Mead became less and less approachable by her intimates and more so by strangers and famous persons. If we imagine a diagram in which groups of Mead's familiars are arranged in concentric circles around her according to their degree of familiarity, we find that the further removed, the more pleasant the contact. Many assistants in the museum tower were afraid of her lashing tongue, and she entirely avoided fellow curators. In midlife, her guiding myths began to crumble. "The central place of the hearth," Mead as "the mother of the world," family as safety net through which one never falls were three of these myths. In her memoir, *With a Daughter's Eyes*, Mary Catherine Bateson pinpoints the problem in a letter her mother wrote in 1955 but did not send. Instead, she left instructions to Marie Eichelberger to send it to a number of relatives and friends in the event of her sudden death. After Mead's gradual death by cancer, Eichelberger passed this letter on to Catherine, in whom it sparked bittersweet feelings. The letter expresses Mead's regret that her life lacks any form of intimacy.

> I have become increasingly conscious of the extent to which my life is becoming segmented, each piece shared with a separate person, even where within the time and space of that segment, I feel that I am being myself, and my whole self in that particular relationship.... Since the break up of my marriage, far less of my life has been shared with one person, and a multitude of special relationships, collaborations, slight gaieties, and partial intensities have taken the place of a marriage which once occupied so much of my time and attention. Distance now separates me from people who once were able to keep most of the threads in their hands.
>
> I prefer a life in which each important feeling and thought can be shared with someone whom one loves, friend or spouse — several friends, teachers and pupils. It has not been by my choice of concealment that anyone of you have [sic] been left in ignorance of some part of my life which would seem, I know, of great importance. Nor has it been from lack of trust — in my person — on my part, but only from the exigencies of the mid-twentieth century when each one of us — at least those of us who are my age — seems fated for a life which is no longer sharable. (Bateson 115–16)

Bateson recognizes the sense of fragmentation in this letter as a common theme in both her parents' lives, and knows that it increased in intensity through the rest of Margaret's. "The note of sorrow Margaret expressed in this draft letter continued and deepened through the remainder of her life as a background to the achievements and satisfactions of the next twenty-five years" (Bateson 116). Optimistic Margaret, who redefined Family so that it would not leave people out, was lonely. She did have friends, but there was something secondary about her friendships. As Grinager and several other writers have commented, she was extremely demanding of her friends. They nearly always served her in her work. She did not have a driver's license, and expected a ride from the airport, on short notice, whenever she landed in the town where some friend or acquaintance lived. She also preferred staying at their homes, unless there would be plenty of friends at the meetings she attended, in which case she could stay at the appointed hotel without fear of loneliness. Her daughter claims that though she spent a great deal of time alone, it was only because of the demands of reading and writing. Otherwise, she could not tolerate it. At the end of a sixteen-hour day at the Mexico conference, a day in which either Pat Grinager or Dana Raphael had trailed her full-time for their celebrity portrait, Margaret called up Dana to say, "why didn't you come back to the room with me? There's nobody here and I feel abandoned" (Grinager 167). Several friends knew that she could not eat alone: "Solitary relaxation was not something she sought and she had a virtual taboo on solitary meals, making every effort to eat with a companion, almost all pleasure defined as social" (Bateson 122).

June Goodfield, intellectual historian and friend, wrote that Mead was "pervaded by deep loneliness," and provides a possible reason — she could never ask for emotional help, only practical help like rides, meals, or secretarial assistance. "[A]lthough she never admitted it, she was very vulnerable too. She cared immensely and gave generously, often to the point of being drained and impatient, but she rarely sought support. Everyone was her neighbor in the Biblical sense, yet she demanded far too much from those close to her" (Howard 361). Because of her busy international lecture schedule, her lack of domesticity, and her temperament, it was not feasible for Margaret to repay these demands in kind. Consequently, all relationships took on an unpleasant one-sidedness, which most of the friends and acquaintances (interviewed by Howard) chose to ignore. Barbara Walters caught a glimpse of Mead's vulnerability in a taxi they shared after an interview. Walters complained that living alone, being a public figure (who is subject to criticism), and trying to raise a daughter was lonely and difficult. To Walters' surprise, Mead concurred. "You, the most independent woman in the world?" replied Walters (Howard 364–5). Perhaps that fearless image cost Mead her closeness with others.

Although she would never consent to undergo psychoanalysis, some members of Mead's extended family ventured their own analyses of her complex personality. Colin Frank, who grew up as Cathy's "brother" in the Perry Street coop, became a psychotherapist and was candid in his interview with biographer Jane Howard. He described the ménage at Perry Street, which began at the same time as World War II, as a terrible time in the world and in his father and Mead's lives. Why was it terrible for Larry Frank and Margaret Mead? Perhaps he is referring to the fact that anthropologists were no longer permitted to go to the (international) field in wartime, and joined the war effort instead. Bateson volunteered, first in his native England (missing his daughter's birth), then in the United States Office of Strategic Services, which assigned him to plan and execute "psychological warfare" in China, Burma, Ceylon, and Thailand. The work was "disastrous" for him, said his wife, and he considered suicide. Margaret worked for the National Committee on Food Habits in Washington, D.C., far removed from her two-year-old daughter. Perhaps, on the other hand, this was a "terrible time" for Colin's father because he was hen-pecked by Margaret. The general tenor of Colin Frank's remarks about Margaret fit well with the epithet she earned in college, which stuck for a lifetime — the "castrator." For instance, "I wanted him [Colin's father] to take more credit; I wished he would say '*I did that*!' the way Margaret did. She was more self-promotional than he was, although not as much as she later became" (239). Although she was not his biological mother, Colin feels a son's ambivalence about Margaret's contributions to his psychological development. He is somewhat glad she encouraged him to be the best he could, but on the other hand, wishes he had not been raised in such a competitive atmosphere: "the aura of Margaret took away from the value of ordinary life ... all of us bit players were there to make up for the sacrifices that she made" (Howard 244). His younger brother, Kevin Frank, also sensed a sad urgency in his parents and their friends. Although energetic, Kevin feels that they were permeated by "sadness—a longing for the formula, the concept, the doctrine, the observation that would be ultimately revealing, which the circular nature of conceptual reasoning makes it impossible to find—a longing to understand, yet perhaps a not complete awareness of why the whole thing seemed so frustrating" (Howard 291). The combination of great hopefulness with stony realism marked Mead's outlook too.

The other relative who psychoanalyzed Mead for the record was Leo Rosten, her sister Priscilla's husband.[6] He looks behind Mead's confident exterior to find (what the Samoa letters to Benedict also reveal)—insecurity about being ignored. "Her elaborate quest for renown ... sprang from something not everyone realized: ...she never had enough recognition, enough affirmation" (Banner 10). Even after *Coming of Age*, she could experience deep insecurities about being a good enough anthropologist. When Ruth Benedict failed to write regularly to her in New Guinea, Margaret thought it was because she was not doing stellar work, and vowed to find new field methods to regain Benedict's good graces, to give the older teacher something to use in class. Benedict reassured Mead that the reverse was true: that Mead was suffering in the field because she longed to work with people "better than herself," and these were not available. In another letter from Bali, Margaret regretted being away from Columbia so long and worried about the effect of her absence on her status with students and colleagues in anthropology: "why isn't anyone asking my advice?" and "why is there a myth that I'm an ogre?" Once again, Ruth assured her. The myth was fueled by colleague's insecurities about their own inefficiency and thoughtlessness: "When you feel you are pouring out your wisdom and sharing it with them, they are feeling sunk because you're so good (your brain, not your good nature)" (Lapsley 255). (By the way, that qualifier says a lot!) Mead needed constant approval and intellectual exchange in order to feel good about herself and her way of doing fieldwork.

Mead once said about her youth, "I never had enough companionship" (Howard xiv). That was despite three siblings, a beloved grandmother, servants and their families, and friends. She vowed never to be without a trio of intimates: a best friend, an admired teacher, and a man. The situation of frustrated social need only got worse as she grew older, and considerably worse once she lost her third husband. Her brother-in-law remarked that the Mead siblings were never close, and maybe that was why Margaret made transferences, which are intense personal feelings that one originally has for a parent or someone close, shifted to another figure later in life (Howard 360). Certainly Ruth Benedict fits that description, being Mead's mentor in the field. But Benedict also loved her as a friend and confided her own deepest feelings to her. Mead sought out other mentors as well — Gregory was one because of his urbanity and his respectable background in biology. She believed his science background could help her to make her own intuitive approach more empirical. Margaret suffered to a certain degree from a syndrome common to women who succeed in traditionally male careers, the feeling that they are frauds who are somehow pulling the wool over the public and the profession's eyes. These women sometimes feel that their success is based on a carefully constructed professional front, while underneath, they are lacking in fundamental skills or native ability. Her brother-in-law believes that, despite her arrogance, "essentially she was modest, and insatiable for reassurance. She grabbed every lecture opportunity, every seminar, every chance she could find for applause" (Howard 381). Mead admitted as much when she said, "Even writing a book is, for me, a performance." Howard maintains that she felt the performance requirement of her profession "literally and always"(381). The thumb-stick and red cape she bore in her later years were costuming for a part — the theatrical role of the eccentric lady anthropologist.

What were the causes for such insecurity? Was there an unresolved primal trauma in Mead's childhood? When Mead was five, her younger sister died, and Mead later believed that this was her primal trauma. Not only had the parents allowed Margaret to name this child (Katherine), but to make matters worse, she had died in their shared bedroom. However, Mead claimed that the birth of a new sister completely resolved any survivor's guilt or other kind of neurosis. At her young age, she believed that the new baby was sent to replace the old, and instantaneously formed "an abiding faith that what was lost would be found again" (BW 63). Although plagued by recurring dreams of dead babies until her mid-thirties, Margaret ceased believing that they represented an unresolved fear that she had somehow killed her younger sister. She thought they might represent a fear for her two younger sisters, Elizabeth and Priscilla, whom she had failed to protect from their father's manipulations. Eventually, she interpreted dead babies in dreams as a part of herself that had been killed in childhood (Banner 264). Could that dead part of herself have been the belief in the goodness of marriage, as she watched her parents argue and hurt each other? Maybe Mead's unresolved trauma was the horror of her parents' marriage. She saw unhappiness in her home life that she was afraid of replicating in her own. In kindergarten, she clung to her teachers' skirt in tears when it was time to go home. Her child's death caused an extended depression in Mead's mother, and her father solaced himself in sexual escapades. In her own life, Margaret desperately wanted to avoid what had happened to her mother.

Emily Fogg Mead was a feminist with a Ph.D. in Sociology who never achieved her goal of a satisfying work life because she had five children and a traditional husband, whose business ventures put the family through financial ups and downs. To add to this feminist's disappointment in life, her husband was a man's man, inexpressive and superficial, who dallied with other women. How did this combination of personalities affect Margaret? On the positive side, being firstborn, intelligent, and confident, Margaret was Daddy's girl. His pet phrase was, "there's no one like Margaret," and he recognized her intellectual superiority to her brother. On the other hand, he could be authoritarian and controlling. He withheld money to try to make Margaret comply with his wishes. He refused to send her to Wellesley, his wife's alma mater, saying she was too delicate for college. Margaret offered to go to a far less prestigious school, Indiana's DePauw University. Choosing his alma mater must have stroked his vanity, since he relented and paid the bill. The next obstacle was graduate school. He withdrew his support for her doctorate at Columbia because she would not give up her fiancé, Luther Cressman. Ruth Benedict stepped in with a no-strings-attached fellowship of $300.00. In short, Margaret learned how to get around men's restrictions on her plans. "I never let a man's wishes stop me from doing what I wanted to do, but once I came close," she said, and that was when Edward Sapir tried to convince her not to go to Samoa. Throughout her life, Margaret must have been determined not to become her mother, a woman whose ambition and happiness were thwarted by a selfish husband.

Like many first-borns, Margaret wanted to please her parents and to follow in the footsteps of their professions, though not their lifestyle. However, between the two of them — both educated, both progressive — her father was the one who projected an image of success and satisfaction. He earned full professorship in the Wharton School of Business at the University of Pennsylvania, and tirelessly promoted adult education, setting up extension campuses throughout the state. Margaret identified with him, in part because he gave her more affection, but largely because he was the more powerful of the two parents, the one who realized his potential. She inherited or adopted many of his character traits, both positive and negative. In this sense, we might ask whether Mead's temperament arose more from nature or from nurture. But the question that she always asked was why she had come out different from her siblings, never whether she resembled her father. On the positive side, he was a good listener who jotted down interesting things people said. He taught Margaret how to make an effective speech, he had a photographic memory, and he was good at languages. He had a missionary's zeal for his life's work, believing that the working class could and should undertake secondary education to improve their lives. On the negative side, he was bullying, demanding, undependable, unfaithful, sarcastic, stubborn, manipulative, aggressive — what Banner calls the "hypermasculine" Victorian male in an era of "imperialism, violence in sports, and the robber baron mentality in business" (75). It is uncanny how many of these traits Margaret developed. She was called all of these things by many of those who knew her. She was also applauded for optimism, generosity, brilliance, energy, loyalty, and hard work.

Her vaunted optimism came at a cost, however, and that was the repression of her negative feelings. Freud taught that what is repressed will inevitably return, whether in neurosis or psychosis. Margaret's deepest neurosis was the fear of being alone, not because she could not care for herself, but because it signified that she was unrecognized or unacknowledged. It is interesting that her best friend, Ruth Benedict, was very good at being alone. In fact, Benedict dealt with her own neurosis by self-analysis and tranquil withdrawal. So expert did she become at calming herself that many people remarked upon her serene detachment. This coping mechanism did not exempt her from full participation in professional and social activities (she was eventually Chair of Columbia's Department of Political Science, in which Anthropology was a division). On the other hand, such detachment suggests its own brand of over-compensation. In comparison to Mead's manic over-involvement, however, Benedict's detachment was easier on those around her.

If Margaret were so able to follow in her father's successful footsteps, then why should she have fears and insecurities, and what were they? Since she recorded dreams, wrote to Ruth as a soul mate, and even divulged a few personal things in her autobiography, we have a substantial record of Margaret's inner life. Two scholars, American historian Lois Banner and New Zealand psychoanalyst Hillary Lapsley, have addressed psychological issues in the Benedict-Mead friendship. Banner writes that Mead had fears of exposure, of disgrace, of professional failure, and of rape. She feared hurting people close to her, being "mannish," and being bisexual. In the beginning of her marriage to Gregory Bateson, she tried to project an image of insecurity so that he would feel more masculine (345). These many fears had psychosomatic symptoms, such as recurring neuritis in her right arm, eye trouble, and ankle breaks and sprains. She also overworked to the point of nervous fatigue. She had several miscarriages and suffered one major depression in New Guinea. She was proud of avoiding depressions, perhaps because her mother was incapacitated by depression after her fourth child was born, forcing Margaret to take over many of the household duties at age eleven. Even without the benefit of depth psychology, it is easy to see that Margaret was driven to succeed in the public sphere by the negative example her mother set by missing her chance in that arena. Combined with many gifts, this strong drive resulted in an excellent and unrivalled career, especially for a woman of her time. Mead's great gifts were enormous energy and vision, complete attunement with the times, the ability to collect and synthesize information rapidly, and compassion for others.

Although children do not always give the most objective analysis of their parents, they can sometimes understand them better than most outsiders. Mary Catherine Bateson was the first commentator to reveal her mother's bisexuality to the public, in her 1984 memoir, *With a Daughter's Eye*. She expresses regret that her mother did not share this information with her during her lifetime. Although understanding the demands of her professional life, Bateson feels that her mother's secrecy cost her intimacy with her child and with others. Mary Catherine remarks that being intimate with both Ruth Benedict and Gregory Bateson "created a kind of isolation, an isolation of secrecy" (118). This isolation was not only Margaret's; Cathy herself feels the loneliness of her growing years, although she is exquisitely careful not to assign blame.

> Whatever it felt like to have both parents progressively withdraw, beginning when I was just over two years old, and then be absent for increasing periods of time, these memories are almost entirely lost, as is the memory of the stress created by introducing an additional child [herself] into an already complex household where a younger infant [Colin Frank] had a right to those years as the center of concern and delight. A sense of aloneness, a sense of an occasional frighteningly inimical response to midnight demands for comfort — these must be dredged from the deepest part of dreams and then combined with the adult knowledge that any caretaker would be exasperated and burdened under those circumstances. I was rich beyond other children and I had the delight of a playmate who was like a brother, Colin, and yet there were all those partings. There were all those beloved people, yet often the people I wanted most were absent. (39)

Catherine has chosen to alleviate her own loneliness, and avoid the loneliness she sees in her mother's life, by staying with one husband, whom she married at the age of twenty-one (127). Colin Frank corroborates Cathy's view that the "mother of the world" was not the best mother of her own child: "She [Margaret] and Cathy had mental intimacy, but Mead was not the cloth mother, as psychologists put it. She had no appetite or skill for day-to-day life. Face it, she couldn't have been the cloth mother and done all the other things she did in the world" (Howard 238). As far as worldly success is concerned,

Cathy came out well. She earned a doctorate in linguistics at Harvard, was Dean of Northern Iran University and Amherst College, taught at numerous universities, is President of the Institute for Intercultural Studies, and lectures around the country. She is a mother and a grandmother, and is still married to her first husband. On the other hand, the social scientist sadness syndrome that the Frank boys diagnosed as a professional hazard, seems to have been inherited. The tone of Cathy's memoir is sad and wise. In addition, she is angry: "without Derek Freeman I would have lacked a spice of anger that has strengthened some of my affirmations" (8). Margaret Mead has left her quite a legacy to defend, and very large shoes to fill.

More than any other locale, Margaret Mead will always be remembered for Samoa. *Coming of Age in Samoa* chimes with youthful optimism: she accomplished her first research mission, finding the Samoans to be good, and exhorting Americans to be more like them. In her life, she followed her own advice, borrowing what she took to be their ethos, though it looks very different in an American manifestation. She agreed to non-monogamous marriages with her first and third husbands; she maintained a deep and occasionally sexual relationship with her original mentor, Ruth Benedict; and she tried at all times to expand her social world, making networks of interdependent friends and colleagues (what one friend called her "maypole conspiracies," with her in the center of each). Despite her praise of the "lack of specialized feeling" among Samoans (since it precluded dangerous attachments), Mead severely missed the intimacy of close attachments when she no longer had any. Throughout her life, she fought many different kinds and sources of isolation. The competitive world of academia isolates professors from the marketplace and from each other. Anthropology isolates its ethnographers, each in his jealously guarded territory. American women were isolated by either of their two choices, marriage and career, especially since having a career was not the social norm for women. American women in male careers were isolated by men's fears of their competence. Fieldwork isolates the observer from his or her country, culture, and community. Homosexuality isolates individuals when social stakes are too high to risk exposure. Fame isolates the famous by demanding time and performance of an expected role. Driven by a fear of isolation,

Mead tried to make the world her family. Her essential loneliness suggests the impossibility of the dream. The difference between Samoa and America is greater than suspected, and this difference affirms her very first principle. All cultures are relative and distinct and must be evaluated within their own framework.

VII

Island Dreams: Pitcairn as Paradigm

> We all hold a place within our hearts — a perfect place — which is in the shape of an island. — Dea Birkett, *Serpent in Paradise*

Pitcairn Island carries the weight of two centuries' worth of psychological projection. Because of the mysterious appeal of the mutiny on the *Bounty*, and of Fletcher Christian's chosen place of exile, this one-by-two-and-a-half mile island in the very middle of the South Seas "fulfilled our longing for perfection like no other before it" (Birkett 15). Writer Dea Birkett sets out in the early 1990s to capture her own dream on Pitcairn, and in the process, discovers that "dreams ... should never be visited" (294). Its very smallness made Pitcairn a prison. Because of the interdependence of its forty residents, emotions had to be suppressed. So did the fact of island murders and individual differences — all in the interests of getting along, and not exploding like the volcano that made the island in the first place.

When Captain William Bligh commandeered HMAV *Bounty* in 1787, his mission was to proceed around Cape Horn to collect breadfruit seedlings in the South Seas. These would be cultivated in the Caribbean in order to feed plantation slaves. Bligh was a tough master, putting men on two-thirds rations early in the voyage, persisting in futile attempts to round the Horn despite violent storms before finally doing an about-face for the Cape of Good Hope, and refusing to land between Van Dieman's Land and Tahiti. After five months of breadfruit collecting and cavorting with Tahitian natives, the *Bounty* men set sail for the West Indies. Within three weeks, master's mate Fletcher Christian led a mutiny. The crew jettisoned Bligh with eighteen loyal crewmen in a twenty-three-foot launch. Miraculously, he survived the 3618-mile journey to Timor in the Dutch East Indies, without navigation charts and with little food. Propelled by his desire to denounce Christian and save his own reputation, he returned post haste to England, where the naval tribune quickly exonerated him and dispatched Captain Edward Edwards to arrest Fletcher Christian and his mutineers.

Thanks to the remoteness and inaccessibility of Pitcairn Island, Christian and crew were never found. Locating a description and chart of Pitcairn in Captain Bligh's cabin, Christian and his crew of nine British and nineteen Polynesians chose this island, out of the myriad other choices, because of its difficult harbors, steep sides, lack of inhabitants, and fresh water streams. Once landed, they burned the *Bounty* so that there would be no obvious clue to their whereabouts. However, now there was no means of escape, since there were too few trees on the island with which to build a new ship. What happened next is uncertain, but Birkett pieces together the following account: the Polynesians who had accompanied the British crew to Pitcairn grew discontent because they were given no stake in the new land, and had to share their women with the white men. Of the twelve Polynesian women, nine were given to white men, which left three to share among the six Polynesian men. The Polynesians then mutinied against the mutineers. Four got away (though hiding out in such terrain defies the imagination); these were called Quintal, McCoy, Young, and Adams. McCoy got drunk and threw himself off a cliff; Young and Adams chopped up Quintal with a hatchet; Young died of asthma or consumption; only Adams remained. Although accounts vary, most contemporary inhabitants claim lineage from the original settlers, and even continue the tradition of hierarchy inherited from those initial shipmates. Carrying the name "Christian" ensures the highest esteem in Adamstown, Pitcairn's one village. However, even island inhabitants disagree about the morality of Fletcher Christian's actions.

This is not a happy story, yet citizens of countless countries have pinned their romantic projections on Pitcairn Island ever since Fletcher Christian landed there. Researching Pitcairn for a full two years before her visit, journalist Dea Birkett found an extraordinary amount of interest worldwide. There are Pitcairn societies, five movies about the mutiny, over two hundred books that interpret the event in their own way, and a Pitcairn newspaper with the highest per capita subscription in the world (five thousand, which is more than one hundred readers for every Pitcairner). Birkett's search also uncovered the inevitable fanatics, including one pathological liar named Tim, who invited her to sail to Pitcairn on his yacht. Tim used his Pitcairn dream as a way to find an identity and a belief system, even though he had never visited the island. Although the Pitcairn enthusiasts attribute high standards of courage, nobility, and justice to the mutineers and their descendants (qualities required to overthrow their Captain and establish their own community), reasonable minds may come to different conclusions. In some ways, an analogy to the American Pilgrims is apposite: they must have been religious, courageous, and individualistic to set sail for unknown shores to start a new life. In addition, they could be intolerant (Salem witch trials), imperialistic (treatment of Native Americans), and proud (God's chosen people). Among many other things, Birkett found that Pitcairners were also intolerant and proud. Perhaps she should have known better than to enter a Seventh Day Adventist island community without considering the threat she posed to them: single, pretty, and intelligent, she disrupted their religion, the security of their marriages, and the sanctity of their reputation abroad — like a latter day Margaret Mead!

Birkett's account of her several months' stay on Pitcairn Island is titled *Serpent in Paradise*. Like all good titles, it has more than one meaning. Who or what is the serpent? She is, insofar as she carries the apple of temptation to islanders, in the forms of her body and her outsider's capacity for judgment. But *they* are also the serpents, insofar as they inhabit paradise in a far from innocent way. Birkett leaves earlier than expected, because she fears they will kill her. It is only when their ostracizing leads to physical menace that she finally succumbs to the signs that she truly does not fit in, and will likely pay the consequences. Incidents pile up and point to an unspoken truth. Her friends ignore her when she falls and suffers a concussion on a boat. She witnesses her neighbor plant nails, points up, in another neighbor's walkway to avenge him for cutting some trees. She notes the unreasonable rate of accidents when the island gathers to shoot breadfruit out of tall trees with .22s. These events bring home a new understanding of the route that suppressed emotions take in a community that can hide itself from the justice system of the outside world. From their viewpoint, she has a cardinal sin to atone for, sleeping with "Nigger," the married policeman whose wife is off-island, while Nigger himself goes blameless. Meanwhile, her landlady's son Dennis alternates between insulting and courting her. She cannot figure out these islanders, prompting her identification with Robert Louis Stevenson's words (which she adopts as an epigraph): "They say it scares a man to be alone. No such thing.... What scares him worst is to be right in the midst of a crowd, and have no guess of what they're driving at."

Although Birkett never directly condemns Pitcairners, her story allows us to do so, according to our own morals. The final incident clinches it for me: Irma, her landlady, asks Dea to carry a letter back to their mutual contact in England, the radio operator Ron, who makes and receives occasional calls to and from Pitcairn. The serpent of Gossip viciously bites its victim when the radio operator reveals the letter's contents to Dea. Irma disapproves of Dea's encounter with Nigger, and wants to let the outside world know about it. This incident suggests a few things about Irma, Pitcairn, and the insecurity of islanders in general. On the one hand, Irma's letter may have been intended to cast doubt on unpleasant truths about the island that Dea might later publish, by damaging her credibility with a Christian audience. On the other hand, the letter suggests a request for affirmation from the wider world, as if the islanders themselves were insecure about their moral soundness. Irma's revelation to Ron seems strangely irrelevant and misdirected, though keenly malicious.

Dea Birkett entertains, amuses, and enlightens with her engaging tale of life among the Pitcairners, but what relation does it bear to my current study of modernists on islands? How might Birkett's account be considered paradigmatic of the other island experiences recounted herein? Darwin was not looking for dream-fulfillment in the Galapagos, and Synge was not disillusioned by his beloved Aran Islanders—quite the opposite. Analogies to the other three innovators are more apposite, as Lawrence's plans for Rananim and the Island schemes were more or less self-conscious exercises in fantasy for psychic relief; Mead's projections of Samoan mores back onto her New York life among the intelligentsia were fraught with tension; and Gauguin's profligacy depleted various islands' tolerance of him — no island could be so "primitive" as not to mind his own savagery. (By the way, I intend no comparison between Birkett's peccadillo and Gauguin's promiscuity.) What all five stories do have in common, however, is the use of islands as imaginative space for outsiders' profoundest intellectual and emotional needs, and Birkett is particularly aware of islands' ideal capacity for psychic projection: "Everything about an island suggests a complete and private world, upon which you can make your mark," she writes (15). Unlike any of the five cultural giants, Birkett enumerates her fantasies one by one, only to disassemble them, while yet retaining compassion for herself and equally for the forty souls who call Pitcairn home. Her story elucidates both the potentials and the pitfalls of island dreams.

A prominent theme of *Serpent in Paradise*—in fact, the reason why the dream fails—is Birkett's inability to relinquish her individuality in order to conform to the island community: "I had acquired most of the skills necessary to live on Pitcairn.... But the core skill to being a Pitcairner — to be part of a whole, not a whole in yourself—still eluded me. I was unable to shake off an unrelenting individuality" (274). Birkett noticed that people were not supposed to make individual decisions on the island, and that every activity was to be performed with others. Whenever she took a walk by herself, her fellow islanders gossiped about her. Because she had gone on a wood-collecting expedition to another island, which involved sleeping on the boat floor along with several men, a rumor arose that she had seduced a married man. Islanders encouraged her to drink coffee, and another outsider to smoke cigarettes, even though they then judged them for breaking Adventist rules.

The "unrelenting individuality" of Birkett is also a core characteristic of the innovators in my study. Iconoclasts, idealists, and individualists, they needed remote island space to cultivate these core qualities, the very existence of which meant that they were (to varying degrees) outsiders to their own societies. It also meant that they would not assimilate well to new, more homogeneous cultures, returning us to the premise discussed in my Introduction, that island dreamers are caught in the double bind of the desire for belonging and the visionary capacity to make significant change. Birkett's lesson — that her own individuality and privacy were indispensable to any formula she might create for personal happiness— is one shared by the other island visitors in this study. In the aforementioned double bind, individuality pulls more strongly than community, at least for individuals with a strongly creative spirit.

D. H. Lawrence's lifelong flirtation with Island schemes that did not materialize aptly illustrates this stronger pull, perhaps because he was more self-aware than the others, or perhaps because his milieu was the modern novel, which took as its theme Romantic Individualism in a nonromantic age. Margaret Mead's shibboleth was Family (meaning community), yet she is perhaps the most self-actualized female individual of the twentieth century. The fact that she could sell the idea of Family publicly while missing it personally suggests that her individuality was too strong to make a successful blend with most true communities. Synge paid the price (of public scorn) for his individual vision of Irish peasantry and its role in the national literary revival. Gauguin needed distance to become different in the overpopulated field of Post Impressionism, and worked at creating his own hero-legend in order to stand out from the crowd. If it is harder to see him than the others as highly individualistic today, perhaps this is because the stereotype of the artist as promiscuous, alcoholic, flamboyant, and misunderstood has become so widespread ... but he had much to do with the origins of the stereotype. Darwin's individuality is in some ways the clearest, in other ways the most muted. An iconoclast, he took on nothing less than the Victorian Establishment, challenging their religious, social, and scientific views. On the other hand he worked collectively, he fit in with his social class, and he delayed his publication until so long after his island experience that the vibrancy of newness and youth were attenuated. He waited until Alfred Russel Wallace had arrived at the same conclusions, making Darwin no longer the lone individual crying in the wilderness, a position he did not relish.

Besides exploring islands' resonance as icons of utopia and individuality, Dea Birkett clearly expresses islands' potential to receive our negative projections as well. She cleverly figures the island as the repository of the unconscious, the source of mystery and fear. Recalling Lawrence's man-who-loved-islands, Birkett takes us on a first-hand journey towards the threats posed by the Other. Did Dennis, for instance, love or hate her? Why were there so many cliff deaths in a place where people's feet had adapted suction-like qualities from their daily mountain climbing? Why did perfectly friendly, hard working island enthusiasts like Perry, Mary, and Alison not fit in? How could she know how to please her landlady when her efforts to help were seen as malicious? How could anyone ever understand anyone else if frank talk was discouraged? She found most Pitcairners' speech to be, to put it euphemistically, "non-discursive." Pastor Rick has enough insight to tell her why they are silent: "They're frightened of retribution. They're scared that if they say something against someone they'll get back at them at some later date. You rely on everyone, every day, for your life — in the longboat, climbing the ladder. You just can't walk away from things here" (282). But Pastor Rick does not have the self-awareness to correct his own hypocrisy: he wears fluorescent pink, see-through hot pants every day, yet condemns dancing and drinking because "they seduce."

Hypocrisy and ambivalence are fearful things, pointing to the unknown in the very heart of life. Birkett leaves her island with emotions as confused as the Pitcairners': "I was scared on Pitcairn, so scared that even when I went out to the duncan [outhouse] I grabbed a broom handle on the way. Yet I felt pain at the thought of departing. I both feared and wanted this island, perhaps as Dennis both feared and wanted me" (287). There is a price to pay for following our dreams, yet Birkett can only have matured from her experience. She writes: "Dreams should be nurtured and elaborated upon; they should never be visited. By going to Pitcairn, I had vanquished the perfect place within myself" (294). The reader is reluctant to believe that Pitcairn destroyed Birkett's spirit along with her innocence, and prefers to think that the experience merely made her a better writer and a wiser person. With the exception of Lawrence's Cathcart, none of the island dreams in this study killed their dreamers, and we easily recognize Cathcart as pathology or science fiction — an allegory rather than a reality. New perfect places will grow within Birkett. The creative spirit is filled with desire; the nature of that desire is to grow where (trans)planted.

Islands are such powerful repositories of psychic activity that they can make us see our own ambivalence and hypocrisy. Darwin and Gauguin came face to face with colonial abuses and their own privilege. Mead had to confront her aloneness in a place that did not sufficiently busy her mind to distract her from her own feelings. Lawrence dreamed his islands as life preservers in the face of illness, marital stress, war, and alienation, yet he knew — via Cathcart — that his own isolation was unhealthy. Synge was the scapegoat for Irish nationalists' troubled ambivalence: if they wished to embrace Irish peasantry for their own nationalist agenda, then he felt they must also accept the brutal facts of lives of poverty. Islands may seem to be an empty slate on which to write our dreams, but they are always already inhabited, like the shores of the unconscious mind. Birkett notes the difference between "land" and "island": "An island, to be perfect, to be paradise, ought to be empty; "uninhabited" is a word attached only to islands, never to land. Land suggests borders, which can be crossed, violated, invaded; but an island, being empty, can be all ours" (14). Her own journey proves how impossible it was to make Pitcairn her own. Far from being empty, it was fully populated, not only physically, but also psychically, with every human challenge. Like the unconscious, islands are far from empty. Even Cathcart's last barren rock held so much psychological content that it undid him. We have seen that islands provide physical conditions for transformational discovery and art. Just as powerfully, they provide imaginative space for creation and destruction, for both dreams and actualization.

Notes and Works Cited

Chapter I

Notes

1. In her essay "Mr. Bennett and Mrs. Brown," Virginia Woolf makes fun of the social-realist style of Edwardian novelists Arnold Bennett, John Galsworthy, and H. G. Wells. To them she contrasts the style of her fellow modernists, who are interested in what a character thinks and feels, rather than in their social status or a lengthy description of their clothing and surroundings.

2. Even though Darwin's work predated the span of literary modernism, we may consider him a proto-modern because he was ahead of his time. His general outlook, one of radically challenging received knowledge, matches the outlook of scientists and artists of a later era.

3. See, for example, Alden T. Vaughan and Virginia Mason Vaughan's book *Shakespeare's Caliban: A Cultural History* (Cambridge: Cambridge UP, 1991).

Works Cited

Arnold, Matthew. "To Marguerite—Continued." *The Norton Anthology of English Literature*, 7th ed. Ed. M. H. Abrams. Vol. 2. New York: Norton, 2000. 1479.

Donne, John. "For Whom the Bell Tolls." From *Devotions on Emergent Occasions. The Longman Anthology of British Literature*. Vol. 1B. Ed. Damrosch and Carroll. New York: Longman, 2003. 1667.

Gauguin, Paul. *Noa Noa: The Tahitian Journal*. New York: Dover, 1985.

Harrex, Syd. "Personal Islands." *Overland* 136 (Spring 1994). 51–56.

King, R. "The Geographical Fascination of Islands." *The Development Process in Small Island States.* Ed. Douglas Lockhart, David Smith and Patrick Schembri. Routledge: London, 13–37.

Lawrence, D. H. "The Man Who Loved Islands." *The Woman Who Rode Away and Other Stories.* Cambridge: Cambridge University Press, 1995.

Nepaulsingh, Colbert. "Islands and Continents." *The Latino Review of Books* 2.2 (Fall 1996). 8–10.

Polack, Fiona. "Writing and Rewriting the Island: Tasmania, Politics, and Contemporary Australian Fiction." *Message in a Bottle: The Literature of Small Islands.* Ed. Laurie Brinklow, Frank Ledwell, and Jane Ledwell. Charlottetown, Prince Edward Island: Institute of Island Studies, 2000. 215–229.

Royle, Stephen A. *A Geography of Islands: Small Island Insularity.* London and New York: Routledge, 2001.

Shakespeare, William. *Richard II. The Complete Works: The Cambridge Edition Text.* Ed. William Aldis Wright. Garden City, NY: Garden City, 1936.

Wallace, Alfred. *Island Life: or the Phenomena and Causes of Insular Faunas and Floras Including a Revision and Attempted Solution of the Problem of Geological Climates.* London: Macmillan, 1892.

Welch, Robert. "'Undoing Time': the Voyage Theme in Irish Literature." *The Classical World and the Mediterranean.* Ed. Giuseppe Serpillo and Donatella Badin. Cagliari: Tema, 1996. 69–76.

Chapter II

Notes

1. The inductive and deductive methods of reasoning tend to overlap in studies of any complexity, which Darwin's certainly was. In the case of botany or zoology, a naturalist often encounters data first, such as the beetle collections that Darwin started as a small boy. Later, he reads textbooks that classify beetles according to some general features. These generalities may be considered the Induction. From such a classification, however, the naturalist may make an imaginative leap to an even broader generalization, then test it in the field. He would then be involved in a deductive move. What matters here is not so much naming the method as observing that Darwin worked intently at both angles–observing and theorizing–and did not stop field work once he arrived home in England.

2. The following passage has been the object of considerable ribaldry: "I was much disappointed in the personal appearance of the women: they

are far inferior in every respect to the men.... The women appear to be in greater want of some becoming costume even than the men" (*Voyage* 350).

Works Cited

Beer, Gillian. "Writing Darwin's Islands: England and the Insular Condition." *Inscribing Science: Scientific Texts and the Materiality of Communication*. Stanford: Stanford University Press, 1998. 119–139.

Browne, Janet. *Charles Darwin: The Power of Place*. Vol. 2 of a Biography. New York: Knopf, 2002.

_____. *Charles Darwin: Voyaging*. Vol. 1 of a Biography. New York: Knopf, 1995

Cox, C. Barry, and Peter D. Moore. *Biogeography: An Ecological and Evolutionary Approach*. Oxford: Blackwell, 1985.

Darwin, Charles. *The Origin of Species by Means of Natural Selection*. 1859. New York: Avenel Books, 1976.

_____. *The Voyage of the Beagle*. 1845. New York: Mentor, 1972.

Leakey, Richard E. Introduction. *The Illustrated Origin of Species*. By Charles Darwin. New York: Hill and Wang, 1979.

Mac Arthur, Robert H., and Edward O. Wilson. *The Theory of Island Biogeography*. Princeton: Princeton University Press, 1967.

Marks, Richard Lee. *Three Men of the Beagle*. New York: Knopf, 1991.

Chapter III

Notes

1. The epigraph is by critic Charles Merki, found in "Gauguin and Tahiti," Christoph Becker, p. 59.

2. The cornerstone of Mathews' argument for physical abuse is that the Gauguins' eldest son told George Crès that he had seen his father "bloody [his] mother's face" *(Paul Gauguin et Vincent Van Gogh*, Victor Merlhès, 1989). Emil did not think that his parents hated each other. On the contrary, he believed that "all his life he respected her profoundly" (*Avant et après* 11). Editing his father's journals reminds Emil of Gauguin's essential goodness: "They bring sharply into focus for me his goodness, his humor, his insurgent spirit, his clarity of vision, his inordinate hatred of hypocrisy and sham" (12).

3. What is noticeably missing from this ideal description, however, is any reference to the typical erotic fantasy of naked women without sexual inhibition which played perhaps a greater role in Paris's fantasy of Gauguin's

retreat than it did in his own description. This is not to say he was not interested in the possibilities, both artistic and sexual, of a land where women went topless and were sexually uninhibited. But the reality was far different. The most common mode of dress for women in public was the missionary dress or Mother Hubbard, a *muumuu* which covered the body from neck to ankle, from shoulder to wrist, with no belting or tailoring to reveal breasts, waist or hips. The traditional *pareu* (brightly colored cotton cloth with white floral designs, manufactured by Europeans) was worn in the home, but often with a jersey (Heerman 151). Indeed, missionaries had so inculcated modesty into Tahitian females that "it had become difficult for photographers to persuade even prostitutes to allow themselves to be photographed wearing only a *pareu*" (Heerman 151). And American artist George Biddle remarked that a man had to make a woman his mistress before she would model nude for him (Eisenman 180).

4. The social principal ruling reproduction in Tahiti was that no child would be denied proper care, so teen pregnancy was not regarded as the problem it is in Western societies (Mathews 180). Whether related to the mother or not, other families simply took children in. The richer families adopted very large numbers of children (Gauguin's hosts, the Tavanas, had twenty-five). There is no evidence that Gauguin ever kept his Polynesian children under his own roof or contributed to their maintenance. Although evidence is inconclusive, Bengt Danielsson speculated that Gauguin's children by Tehamana and Pahura were cared for by their mothers (when Gauguin left them) or adopted. The child by Vaeoho was taken care of by her well-established family (Mathews 242).

5. Whether the sale of eleven out of forty paintings at the Durand-Ruel exhibition constituted a financial success is a matter of interpretation. Mathews says it was (197); Becker, that it was not (79). Perhaps that is because Mathews is comparing his net wealth in late 1893 (10,000 francs from the show plus 13,000 inherited from his Uncle Isidore) to his previous net worth, which had not been this great since a stock market windfall in 1879 (197). Mathews also argues that Gauguin consistently represented himself—to Mette, friends, creditors, and dealers as poorer than he really was, so that they would extend credit, give money, or pay his bills.

6. Gauguin identified with Victor Hugo's hero, Jean Valjean (*Les Misérables*, 1862), painting the character's likeness as an icon in the background of his own self-portrait (*Self-Portrait: "Les Misérables,"* 1888). He donated the canvas to van Gogh, explaining that he [Gauguin] was "like Valjean with his nobility and sweet interior" (Eisenman 122). This is an interesting self-image. Jean Valjean lived in the sewers, convents and jails of Paris while maintaining a Christ-like equanimity towards his fellow humans. In 1888,

when he made the comparison, Gauguin had lived often as a non-paying guest in friends' apartments, while steadily losing faith in his fellow humans.

7. Gauguin's irascible temper was another reason he was companionless. Roderic O'Conor, an Irish painter who met Gauguin in Pont Aven (1894), catered to Gauguin's vulgar tastes by carousing and tale-swapping, but ultimately found him "too offensive," according to Mathews, to follow to Tahiti: "No, but do you see me going to the South Seas with that character!" (203–4) Another artist who considered accompanying Gauguin was van Gogh. When the "*atelier* of the tropics" was slated for Madagascar, he invited himself along, but since this was subsequent to the Arles cohabitation, it is unlikely that he believed Gauguin would accept.

8. One critic assumed that the slightly raised hips of the girl and the strip of yellow paint emphasizing the curve of the buttock were intended to invite rear entry. I believe the hips may appear raised because of the problem of perspective I discuss in this chapter: the feet are sliding off the bed. If Gauguin were interested in rear entry, however, that does not justify the art historian's condemnation, even if s/he considers his or her own sexuality "pure."

9. By the time of his death, he was drinking half a liter of cognac a day.

10. Thadée Natanson wrote that although the figure clusters would make some sense had they constituted individual canvases, the canvas as a whole made no sense: "The composition, in order to justify itself, since it does not do so materially enough, calls for comment; it remains obscure" (in Eisenman 140). André Fontainas was harsher: "Nothing resides in such canvases save the evidence of deplorable errors, for abstractions are not communicated by concrete images if ... they have not first taken shape in some material allegory that ... signifies them" (140).

Works Cited

Becker, Christoph. "Gauguin and Tahiti." *Paul Gauguin — Tahiti*. Stuttgart: Verlag Gerd Hatje, 1998. 1–84.

Eisenman, Stephen. *Gauguin's Skirt*. London: Thames and Hudson, 1997.

Gauguin, Emil. Preface. *Paul Gauguin's Intimate Journals*. Trans. Van Wyck Brooks. New York: Crown Publishers, 1936. 11–12. Trans. of *Avant et après*.

Gauguin, Paul. *Noa Noa*. 1885. Trans. O. F. Theis. New York: Dover, 1985.

Goldwater, Robert. *Paul Gauguin*. New York: Harry N. Abrams, 1983.

Heerman, Ingrid. "Gauguin's Tahiti —Ethnological Considerations." *Paul Gauguin — Tahiti*. 147–165.

Kelder, Diane. *The Great Book of Post Impressionism*. New York: Abbeville Press, 1986.

Loti, Pierre. *The Marriage of Loti*. 1880. Trans. Wright and Eleanor Frierson. Honolulu: University Press of Hawaii, 1976.

Mathews, Nancy Mowll. *Paul Gauguin: An Erotic Life*. New Haven: Yale University Press, 2001.

"Synthetism." *The Oxford Companion to Twentieth-Century Art*. Ed. Harold Osborne. Oxford: Oxford University Press, 1981.

Chapter IV

Notes

1. The term *Protestant Ascendancy* refers to the privilege and power held by Protestants in Ireland for two centuries, after they defeated Irish Jacobites at the Battle of the Boyne and at Aughrim and Limerick in the early 1690s. Used by nationalists and postcolonialists, it symbolizes the domination of a class defined by religion and social status. It is sometimes almost synonymous with the term "West Briton"—an Irish person who sympathizes with England—for the animus it carries. Historian R. F. Foster complicates the term by explaining that the Ascendants did not always feel their ascendancy: "Given this threat [of Catholic challenge to land settlements] the Irish Protestant sense of heartfelt deliverance after Aughrim was inseparably linked to an equally intense feeling of insecurity" (134).

2. Although he made brief visits to the other two Aran Islands, Synge spent the majority of his time on Inishmaan. The people of Inishmore, the big island, were too involved with commerce because they had the largest port. The people of Inisheer, though appealingly humorous, are class-conscious. Although Synge observes class distinctions in Inishmaan as well, he believes that, surprisingly, they have not corrupted the people, "among whom there is still absolute equality" (*The Aran Islands* 92).

3. A curragh is a light, canoe-like boat accommodating two to four rowers. With no keel, it can easily be landed on a pebbly beach. Its lightness and small size make it risky for the large waves common to the Atlantic, however, and many islanders die en route between the islands or fishing at sea.

4. The Blasket Islands are a series of small, mountainous islands off the Dingle peninsula in County Kerry, constituting the westernmost part of Ireland.

5. Synge was often described by contemporaries as reserved. Yeats called him a "drifting, silent man, full of hidden passion" (qtd. in Kiely 30). Biographer McCormack comments that "aloofness had been so long sustained as to have become second nature," but credits his ability to mix easily with people of different classes to this same aloofness (13).

6. Pierre Loti (1850–1923) was a French naval officer who turned his foreign travel to use in a prolific outpouring of part-autobiographical, part-romance novels set in countries as diverse and far-flung as Tahiti, Brittany, Japan, Turkey, India, and China. His account of Tahiti in *Le Mariage de Loti* (1880) inspired both Gauguin and Synge to travel to exotic places, but Synge soon repudiated the sentimentality of Loti's style, in which he perhaps recognized the maudlin, morbid tone of his own early plays. Anatole le Braz was a better model. Synge met him when he lectured on Celtic Mythology at the Sorbonne in 1897, reviewed his work in a Dublin newspaper, and of course read with studious interest his accounts of life among the Celtic peasants of Brittany. Declan Kiberd calls attention to a passage in le Braz's work that he believes motivated Synge to write in the dramatic genre about Aran. Le Braz asserts that the only Celtic people who have not yet produced a national drama are the Irish, because they had suffered too much from real drama to feel the need to create fictional tragedy (Kiberd, *Synge and the Irish Language* 102). This left the door, and the method (of collecting the dramatic tales the peasants tell) open for Synge.

7. These seven centuries of English domination start tolling from the Norman Invasion of Ireland in the late twelfth century.

8. This theater was later, and probably better known, as The Abbey Theatre. It changed names in 1904, when English impresario Annie Horniman offered money to purchase the old Mechanics' Institute and an adjoining morgue for reconstruction as the new home of the Irish National Theatre Society. Working in a morgue, Yeats had to overcome his fear of ghosts, particularly when a skeleton was found in the premises.

9. Member of Parliament and head of the Irish Party, Charles Stewart Parnell (1846–91) was probably the most efficient leader for Irish nationhood before Michael Collins and Eamon Devalera achieved that goal in 1922. At least, the hero cult created by the Irish Literary Renaissance made it seem so. His strength lay in uniting competing nationalist factions on more issues than the central one of Home Rule. He also garnered the Catholic Church's support behind his cause (Fitzpatrick 181–2). Ireland lost the gains he had made for Home Rule, however, when Gladstone threatened to resign as Liberal leader if Parnell did not resign as head of the Irish Party. Ostensibly, Parnell's moral unsuitability was the cause of his dismissal. He was living and had children with Katherine O'Shea, who was married to a previous parliamentarian, Captain William O'Shea. His personality might have been the real cause. Historian David Fitzpatrick describes his leadership style as a "capricious blend of protracted inactivity interrupted by dictatorial intervention" (184).

10. Ann Saddlemyer supplies the word "virile," while W. J. McCormack supplies the hypothesis of virginity. Saddlemyer writes: "He was a strong,

well-built man, muscular, with broad shoulders, standing five feet eight or nine inches tall and carrying his large, finely shaped head upright.... The nervous and virile face was pale beneath his summer tan, the cheeks rather drawn through constant illness; it was not a handsome face, but singularly expressive and sometimes almost sad. His light hazel eyes, at once smoky and kindling, gazed directly and frankly"(ix). Judge for yourself whether this is a handsome face.

11. Keening is an old Irish practice of wailing lamentation for the dead. Both men and women keen, but Synge observes far more women doing it. Similar mourning calls are found in other nations, and Synge compared it to the war-song of the Somalis at the International Exhibition in Dublin (1907). Keening has disappeared from the islands now, and had disappeared from mainland Ireland at the time of Synge's visits to the Aran Islands, emphasizing again the fact that their geographical isolation allowed them to retain practices from earlier times that had been abandoned on the mainland.

12. Mikhail Bakhtin, a Russian literary theorist and critic who combined Formalist and Marxist approaches in the 1920s and 30s, developed a discussion of "Carnival" which can easily be applied to the topsy-turvy moral universe in Synge's plays. Bakhtin associates qualities of carnivals to certain genres of literature, and to the way that literature acts on the reader. In a carnival, for instance, hierarchies are turned on their heads, opposites are mingled, and the sacred is profane; "everything authoritative, rigid or serious is subverted, loosened and mocked" (Selden and Widdowson 40).

Works Cited

Battaglia, Rosemarie. "The Tinker's Wedding." *A J. M. Synge Literary Companion*. Ed. Edward A. Kopper, Jr. New York: Greenwood, 1988. 61–68.

Fallis, Richard. "Art as Collaboration: Literary Influences on J. M. Synge." *A J. M. Synge Literary Companion*. 145–160.

Fitzpatrick, David. "Ireland Since 1879." *The Oxford History of Ireland*. Ed. R. F. Foster. Oxford: Oxford University Press, 1989. 174–229.

Foster, R. F. "Ascendancy and Union." *The Oxford History of Ireland*. 134–173.

Grene, Nicholas. *Synge: A Critical Study of the Plays*. Totowa, NJ: Rowman and Littlefield, 1975.

Kiberd, Declan. *Inventing Ireland: The Literature of the Modern Nation*. London: Vintage, 1995.

_____. *Synge and the Irish Language*. Totowa, NJ: Rowman & Littlefield, 1979.

Kiely, David M. *John Millington Synge: A Biography*. New York: St. Martin's Press, 1994

Kopper, Edward A. "Toward an Assessment." *A J. M. Synge Literary Companion*. 211–230.
McCarthy, Patrick A. "Synge and the Irish Literary Renaissance. *A J. M. Synge Literary Companion*. 161–172.
McCormack, W. J. *Fool of the Family: A Life of J. M. Synge*. London: Weidenfeld & Nicolson, 2000.
Morrissey, Thomas. "Prose." *A J. M. Synge Literary Companion*. 15–24.
Price, Alan. *Synge and Anglo-Irish Drama*. London: Methuen, 1961.
Roche, Anthony. "J. M. Synge: Christianity versus Paganism." *A J. M. Synge Literary Companion*. 107–134.
Saddlemyer, Ann. "Introduction." *The Collected Letters of John Millington Synge*, Vol. I. Ed. Ann Saddlemyer. Oxford: Clarendon Press, 1983.
Selden, Raman and Peter Widdowson. *A Reader's Guide to Contemporary Literary Theory*. Lexington: The University Press of Kentucky, 1993.
Synge, John Millington. *The Aran Islands*. Ed. Tim Robinson. London: Penguin, 1992.
_____. "In Wicklow, West Kerry, and the Congested Districts." *The Complete Works of John M. Synge*. New York: Random House, 1935.
_____. *The Playboy of the Western World and Other Plays*. Ed. Ann Saddlemyer. Oxford: Oxford University Press, 1995.

Chapter V

Notes

1. Michelucci writes that "individual relationships between the sexes" form the core of community (318). Even if sex is only a partial solution to the problem of finding good community, it is a beginning, for modern man needs to locate his "roots and values" within himself "through a profound, almost religious respect for the Self, the Other and the Environment" (318). The question of the difficulty of finding personal relations in a modern world reminds me of W. H. Auden, who said that human beings have two basic social needs: one for community, the other for intimacy. He believed that in modernity, man's primal need for community was being frustrated, and that many men like himself were unconsciously trying to fulfil two needs at once, through having homosexual relations.

2. George Zytaruk claims that the friendship between Lawrence and Koteliansky was so significant and solid, compared to all others Lawrence had, that it constituted Rananim without need for a physical place or any other members (*The Quest for Rananim xxxvi*).

3. For a detailed study of Lawrence's belief in resurrection in the body,

see my book, *Revisionist Resurrection Mythologies: A Study of D. H. Lawrence's Italian Works* (New York: Lang, 1994).

4. Professor George Zytaruk has a somewhat different idea of the cause of the early failure of Rananim. He dates it to Lawrence's first visit to Cambridge University to meet Russell's friends. Lawrence loathed John Maynard Keynes' homosexuality. After the visit, he had recurring nightmares of beetles and psychosomatic symptoms. Zytaruk thinks it was this confrontation with his own ambivalence about homosexuality that caused Lawrence to rethink the very foundation of his utopia, which was marriage between men and women. He notes the language of Lawrence's letters changing after this visit; although he did not give up the vision, he did perforce give up the idea that Rananim would be formed from Russell's inner Cambridge circle. See "Rananim: D. H. Lawrence's Failed Utopia" in *The Spirit of D. H. Lawrence*, ed. Gamini Salgado and G. K. Das. Totowa, NJ: Barnes & Noble, 1988. 266–294.

5. John Worthen and John Turner, in their essay "Ideas of Community: Lawrence and 'Rananim,'" take pains to differentiate the Rananim project from later manifestations of his utopian ideals, which Lawrence calls either the "Island scheme," or just a call to friends to live with him. Turner and Worthen note that many critics call all of Lawrence's schemes by the name Rananim; they believe it is important to make a distinction between Rananim and other forms of cooperative living, for three reasons. First, we need to put Lawrence's utopian scheme in the context of his time, and look at such a longing as "a widespread cultural desire" (168). Second, we must be careful not to judge Lawrence's psychology or politics too harshly, because doing so tends to make psychosis where there are only passing moods, and to oversimplify complex political trends which continue up to the present. Third, we should distinguish the various "Island" schemes according to their differing degrees of viability and the intention behind them. I think Worthen and Turner are trying to rescue Lawrence from a critical perception of fanaticism for persisting in attempts to create Rananim when it was clearly impossible. Psychoanalysist Ernest Jones, whom Lawrence invited to participate, psychoanalyzed the desire for Rananim in no uncertain terms: Lawrence suffered from a God Complex (megalomania). He needed disciples for a New World of his own making, which was an unconscious attempt to resolve his Oedipal complex (159). As I discuss later in this chapter, however, Lawrence was conscious of his complex, parodied it in "The Man Who Loved Islands," and was mature and understanding when people opted not to follow him, which was most of the time.

6. Frieda did not confide in Ottoline like the others, because she was jealous of Ottoline's intimacy with Lawrence. Her fears may be justified when we consider the novel *Women in Love*, in which Ottoline appears as Hermione Roddice, with whom the protagonist has an affair.

7. David Ellis speculates a third reason Frieda tolerated Brett's presence for so long: she was just too lazy to bother to insist that she leave! (242–3)

8. The biographical information on Dorothy Brett comes from the Harry Ransom Center web site: *http://www.utexas.edu/research/fa/brett.dorothy. html.*

9. For a thorough discussion of the similarities and differences between MacKenzie and Cathcart, the story's protagonist, see "D. H. Lawrence's 'The Man Who Loved Islands' in the Light of Compton Mackenzie's Memoirs," by Viktor Link.

10. Jacques Lacan (1901–1981) was a French psychoanalyst who became central to literary studies in the 1960s. His most fundamental modification on Freud's analysis of psychic development is to shift the description of the unconscious. Freud described the unconscious as a pit of snakes, a bubbling cauldron of erotic and aggressive impulses. Lacan describes the unconscious as a realm structured much like language. In turn, he views the stage when the infant begins to acquire language as determinative of the adult behavior and personality. It is this "mirror stage" which concerns us in a psychoanalytic study of Cathcart's psychosis.

11. Donald W. Winnicott (1896–1971) is a British psychoanalyst whose work impacted the Object Relations school of psychoanalysis. This school theorizes that the mother-infant dyad is the primary relationship, one that establishes the ways in which the individual will pursue and interpret his or her relationships for the rest of life. Similar to the concept of the benign universe which critic John Turner uses in his analysis of Cathcart is the concept of "good enough care" by the mother. Good enough care would be the kind of attentions that allow the "true self" of the infant to emerge. The true self feels creative and spontaneous, whereas the "false self" is built on the basis of compliance. Satisfactory care, according to Winnicott, includes both physical and psychological "holding," and — something that an increasing number of homes do not have — mother, father, and infant living together. <*http://www.sonoma.edu/users/d/daniels.objectrelations.html*>

12. As Freud said in *The Problem of Anxiety*, isolation is a way of preventing angry, aggressive, or erotic impulses from gaining expression (Willbern 241, n21).

Works Cited

Ellis, David. *D. H. Lawrence: Dying Game 1922–1930*. Cambridge: Cambridge University Press, 1998.

Kinkead-Weekes, Mark. *D.H. Lawrence: Triumph to Exile 1912–1922*. Cambridge: Cambridge University Press, 1996.

Lawrence, D. H. *Fantasia of the Unconscious and Psychoanalysis and the Unconscious*. New York: Viking Penguin, 1971.

_____. *Lady Chatterley's Lover*. Cambridge: Cambridge University Press, 1993.

_____. *The Letters of D. H. Lawrence*. Ed. James T. Boulton. Vols. ii–vii. Cambridge: Cambridge University Press, 1981–1989.

_____. "The Man Who Loved Islands." *The Woman Who Rode Away and Other Stories*. Ed. Dieter Mehl and Christa Jansohn. London: Penguin, 1996.

_____. *Sea and Sardinia*. Ed. Mara Kalnins. Intro. Jill Franks. London: Penguin, 1999.

_____. "We Need One Another." *Phoenix: The Posthumous Papers of D. H. Lawrence*. Ed. Edward D. McDonald. New York: Viking, 1936. 188–195.

Link, Viktor. "D.H. Lawrence's 'The Man Who Loved Islands' in the Light of Compton Mackenzie's Memoirs." *The D. H. Lawrence Review* 15 (1–2) (1982): 77–86.

Michelucci, Stefania. "A Man Who Loved Islands: D.H. Lawrence and the Paradox of Rananim." *Vite di utopia*. Ed. Vita Fortunati and Paola Spinozzi. Ravenna, Italy: Longo, 2000. 311–319.

Stoltzfus, Ben. "'The Man Who Loved Islands': A Lacanian Reading." *The D. H. Lawrence Review* 29(3) (2000): 27–38.

Turner, John F. "The Capacity to be Alone and Its Failure in D. H. Lawrence's 'The Man Who Loved Islands.'" *D. H. Lawrence Review* 16(3) (1983): 259–289.

_____ and John Worthen. "Ideas of Community: Lawrence and 'Rananim.'" *D. H. Lawrence Studies* Vol. 8. Ed. Michael Bell, Chong-wha Chung, Earl Ingersoll, Nak-chung Paik. Seoul:. D. H. Lawrence Society of Korea, 1999. 135–171.

Willbern, David. "Malice in Paradise: Isolation and Projection in 'The Man Who Loved Islands.'" *The D. H. Lawrence Review* 10 (1977): 223–41.

Zytaruk, George. *The Quest for Rananim: D. H. Lawrence's Letters to S. S. Koteliansky 1914 to 1930*. Montreal: McGill-Queen's University Press, 1970.

Chapter VI

Notes

1. Contrary to this dramatic depiction, there were other sources of teaching before her fieldwork. She had taken a methods course. She had helped her mother in her doctoral research by interviewing Italian immigrants in Hammonton, New Jersey. She had documented her own sisters' childhoods, a family practice that she carried forward when her own daughter was born.

2. Holmes enumerates several re-studies that commented upon earlier

ethnographers' work, such as Kroeber and the Seri people, Li An-Che and the Zunis, Emeneau and the Toda, and Lewis and the Tepoztlan. None of these re-studies, however, has the explicit purpose of systematically evaluating their predecessors' methods and conclusions (14).

3. In this characterization of Tcherkézoff, Freeman is not above the witty nationalistic slur, though he has accused others of such a prejudice against him.

4. Had it not been for Reo Fortune and Gregory Bateson, her second and third husbands, she would not have brought home from the field the curios that she did. She had no passion for objects.

5. She earned this sobriquet in the 1950s, but people had noticed her mothering quality long before then. The Ash Can Cats accepted her as their mother. Her first husband remarked that total strangers would come up to her and say that she reminded them of their mothers. She had a formula for befriending people that could just as well be the credo of a good mother: 1) be useful to them; 2) be amusing for them; and 3) build them up in their own estimation (Howard 27). As Howard remarks, "the world was her family," but with the role of *materfamilias* went both the good (the caring) and the bad (the interfering): "she trusted she knew what was good for people, often better than they might know for themselves" (xv).

6. Priscilla Mead Rosten killed herself after her divorce from Leo Rosten in 1959. She had always been "the pretty one" of the sisters, but not career-oriented. Margaret often wondered why she and her siblings were so different. This difference influenced her espousal of cultural rather than biological determinism. Priscilla's husband believed that the Mead sisters suffered from a lack of closeness, and indeed, Mead seemed more upset by her brother's death from cancer in 1975, than her sister's suicide (Howard 359).

Works Cited

Banner, Lois. *Intertwined Lives: Margaret Mead, Ruth Benedict, and Their Circle*. New York: Knopf, 2003.

Bateson, Mary Catherine. *With a Daugher's Eye: A Memoir of Margaret Mead and Gregory Bateson*. New York: William Morrow, 1984.

Campbell, Donald. "The Mutual Methodological Relevance of Anthropology and Psychology." *Psychological Anthropology*. Ed. F. L. K. Hsu. Cambridge, MA: Schenkman, 1972. 435–63.

Freeman, Derek. *Margaret Mead and Samoa: The Making and Unmaking of an Anthropological Myth*. Cambridge, MA: Harvard University Press, 1983.

Grinager, Patricia. *Uncommon Lives: My Lifelong Friendship with Margaret Mead*. Lanham, MD: Rowman & Littlefield, 1999.

Holmes, Lowell D. *Quest for the Real Samoa: The Mead/Freeman Controversy and Beyond*. South Hadley, MA: Bergin & Garvey, 1987.
Howard, Jane. *Margaret Mead: A Life*. New York: Ballantine Books, 1984
Keesing, Felix and Marie. *Elite Communication in Samoa*. Palo Alto, CA: Stanford University Press, 1956.
Lapsley, Hilary. *Margaret Mead and Ruth Benedict: The Kinship of Women*. Amherst: University of Massachusetts Press, 1999.
Mead, Margaret. *Blackberry Winter: My Earlier Years*. New York: Simon & Schuster, 1972.
_____. *Coming of Age in Samoa*. New York: William Morrow, 1928.
Tcherkézoff, Serge. "Samoa again: on 'Durkheimian bees,' Freemanian passions and Fa'amu's 'confession.'" *Journal of the Polynesian Society* 110.4 (2001): 431–436.
Wendt, Albert. "Three Faces of Samoa: Mead's, Freeman's, and Wendt's." *Pacific Islands Monthly* (1983 April 10–12): 14, 69.

Chapter VII

Work Cited

Birkett, Dea. *Serpent in Paradise*. New York: Anchor Books, 1997.

Index